Applying Nature's Design

Issues, Cases, and Methods in Biodiversity Conservation

Issues, Cases, and Methods in Biodiversity Conservation

Mary C. Pearl, Series Editor
Christine Padoch and Douglas Daly, Series Advisers

This series combines two earlier Columbia University Press series:
Methods and Cases in Conservation Science
and
Perspectives in Biological Diversity.

Books in the Series

Thomas K. Rudel and Bruce Horowitz, *Tropical Deforestation: Small Farmers and Land Clearing in the Ecuadorian Amazon*, 1993.

Holmes Rolston III, *Conserving Natural Value*, 1994.

Joel Berger and Carol Cunningham, *Bison: Mating and Conservation in Small Populations*, 1994.

Jonathan D. Ballou, Michael Gilpin, and Thomas J. Foose, eds., *Population Management for Survival and Recovery: Analytical Methods and Strategies in Small Population Conservation*, 1995.

Susan K. Jacobson, ed., *Conserving Wildlife: International Education and Communication Approaches*, 1995.

Gordon Macmillan, *At the End of the Rainbow? Gold, Land, and People in the Brazilian Amazon*, 1995.

David S. Wilkie and John T. Finn, *Remote Sensing Imagery for Natural Resources Management: A First Time User's Guide*, 1996.

Luigi Boitani and Todd K. Fuller, eds., *Research Techniques in Animal Ecology: Controversies and Consequences*, 2000.

Harold Brookfield, *Exploring Agrodiversity*, 2001.

Applying Nature's Design

*Corridors as a Strategy
for Biodiversity Conservation*

Anthony B. Anderson and
Clinton N. Jenkins

COLUMBIA UNIVERSITY PRESS   NEW YORK

COLUMBIA UNIVERSITY PRESS
Publishers Since 1893
New York   Chichester, West Sussex
Copyright © 2006 Columbia University Press

Library of Congress Cataloging-in-Publication Data

Anderson, Anthony B. (Anthony Bennett), 1950–
  Applying nature's design : corridors as a strategy for biodiversity conservation /
Anthony Anderson and Clinton N. Jenkins.
    p.   cm. — (Issues, cases, and methods in biodiversity conservation)
  Includes bibliographical references (p.   ).
  ISBN 0–231–13410–X (cloth) — ISBN 0–231–13411–8 (pbk.)
1. Corridors (Ecology) I. Jenkins, Clinton N. II. Title. III. Series.
QH541.15.C67A53   2005
333.95'16 — dc22

                                                                2005041252

Columbia University Press books are printed on
permanent and durable acid-free paper.
Printed in the United States of America

c 10 9 8 7 6 5 4 3 2 1
p 10 9 8 7 6 5 4 3 2 1

Contents

Figures

Acknowledgments

This book has benefited from the collaboration of numerous colleagues at diverse institutions. We are especially grateful to Sophia Bickford, University of Maryland, who assisted in reviewing the literature and text, and also in preparing some of the case studies. Other colleagues who played key roles in preparing the case studies are Richard Hilsenbeck, The Nature Conservancy (USA); John Morrison, Jennifer Reed, and Zach Stevenson of the World Wildlife Fund (USA); Cede Prudente of WWF-Malaysia; Denise Rambaldi of the Golden Lion Tamarin Association (Brazil); and Bram Vreugdenhil of the Veluwe regional government (the Netherlands).

The following people provided helpful information: Richard Carroll, Dominick DellaSala, Lou Ann Dietz, Andre Kamdem, Henry van der Linde, Tony Mokombo, Kate Newman, Judy Oglethorpe, Susan Palmenteri, Mingma Sherpa, and Diane Wood of the World Wildlife Fund (USA); Rafael Calderon, Randall Curtis, and Diego Lynch of The Nature Conservancy (USA and Costa Rica); Richard Cowling, University of Port Elizabeth (South Africa); James Dietz, University of Maryland (USA); Andy Eller, Fish and Wildlife Service (USA); Jeff Gailus and Marcy Mahr, Yellowstone to Yukon Conservation Initiative (USA); Júlio Gonchorowsky, National Institute for Environment and Renewable Natural Resources (Brazil); Doug Graham, World Bank; Ted Hass, Bureau of Land Management (USA); Tom Hoctor, University of Florida (USA); Bart Johnson and Amy Vetter, Wildlife Conservation Society (USA); Bob Pressey and Dave Scotts, National Park and Wildlife Service (Australia); and Amanda Younge, World Wildlife Fund (South Africa).

Finally, we would like to thank those who reviewed and/or made specific contributions to chapters 1–3: Robin Abell, Eric Dinerstein, Curt Freese, Meghan McKnight, John Morrison, Judy Oglethorpe, George Powell, Gautham Rao, Doreen Robinson, Holly Strand, and Eric Wikramanayake of the World Wildlife Fund (USA); John Hanks and Karl Morrison, Conservation International (Southern and Central Africa and USA); Anup Joshi, University of Minnesota (USA); and Cláudio Pádua, Institute of Ecological Research (Brazil).

1 Introduction

> Now, rather than human development occurring in a matrix of natural landscape, natural areas occur in a matrix of human-dominated landscape.
>
> — Harris and Scheck (1991:189)

Today's biodiversity crisis is the direct result of the conversion and loss of natural habitat occurring worldwide at unprecedented rates and scales. Between 1945 and 1990, about 20 million square kilometers — or nearly 17 percent of the Earth's vegetated area — became degraded (WRI 1992: 112). Logging and conversion have shrunk global forest cover by at least 20 percent, and some forest ecosystems — such as the dry tropical forests — are virtually gone (UNDP/UNEP/World Bank/WRI 2000). Over half of the world's coral reefs are under serious stress resulting from destructive fishing practices, pollution, and global warming (Hughes et al. 2003).

Many of the forces driving habitat loss continue to increase and, more troubling still, are interacting synergistically — thereby accelerating ecosystem change (Vitousek et al. 1997). For example, logging not only degrades tropical forest ecosystems but also increases the flammability of entire landscapes, leading to further forest degradation (Nepstad et al. 1999). Likewise, scientists expect global warming to result in widespread habitat modification, and widely believe that its effects on polar ice caps will accelerate climate change — thereby contributing to further habitat loss (IPCC 2001).

The process of habitat loss rarely involves outright conversion of natural habitats over entire landscapes, although current technologies make this increasingly possible. Instead, habitat loss is generally a process of fragmentation, and species disappear as the once-intact habitats that supported them become increasingly fragmented.

Habitat fragmentation is a major driver of today's biodiversity crisis. Defined as the conversion of large, continuous areas of habitat to smaller blocks that are separate from one another, fragmentation isolates plant and animal populations that were once continuous over larger areas. With smaller numbers limited to habitat fragments, such populations are more susceptible to extinction resulting from factors such as inbreeding depression[1] or environmental fluctuations. As a result, they have been aptly characterized as "living dead" (Janzen 1986). The process of fragmentation is extensive, and much of the world's remaining natural habitat occurs in fragmented blocks (Gascon et al. 2000) (fig. 1.1). Among its deleterious effects, fragmentation may lead to:

- elimination or dangerous reduction of populations of large, wide-ranging species, including many top carnivores;[2]
- unraveling of entire biological communities — as, for example, when the decline of top carnivores in fragmented habitats results in the "release" of smaller predators and herbivores, leading to overpredation or overgrazing that may eventually eliminate species or destabilize communities;
- destruction or degradation of remaining habitats through the intrusion of edge effects such as altered microclimate or invasive species; and
- disruption of key ecological processes dependent on increasingly rare animal agents — such as pollination, seed dispersal, predator–prey interactions, and nutrient cycling.[3]

Because these ecological processes are critical for the long-term health of biological communities and ecosystems, conserving biodiversity and key ecological services requires stabilizing and, when possible, reversing fragmentation.

Landscape Connectivity and Corridors

Maintaining or increasing connectivity is the obvious solution to fragmentation. A variety of strategies for doing this exist. At one extreme, we could manage the entire landscape to promote connectivity for species, biological communities, and ecological processes. This relatively long-term approach is especially relevant for wide-ranging species that utilize diverse habitats in the landscape. It involves various strategies, such as establishing linkages between existing protected areas and maintaining key connecting

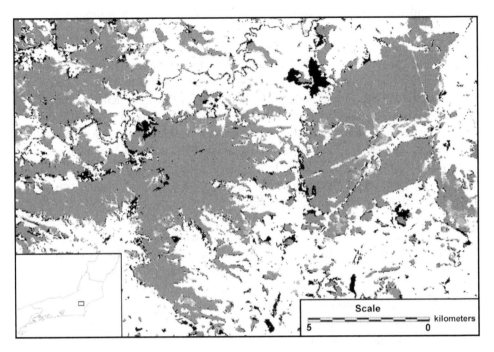

FIGURE 1.1

Effects of habitat fragmentation in an area of Atlantic Forest, Brazil. This figure shows two juxtaposed satellite images from 1988 and 2000. Dark gray represents forested areas that have remained unchanged during this period. White represents deforested areas (primarily pastures) that have remained unchanged. Light gray represents forest regrowth, and black represents forest loss. Note that most forest regrowth took place inside large forest fragments, whereas most forest loss took place in small fragments and around the edges of large fragments. The forest loss was primarily due to dieback from exposure to more extreme microclimate and fires, rather than forest clearing.

features such as riverine systems. At the other extreme, we could manage specific habitats for connectivity. This would typically have a more focused, short-term objective of protecting less tolerant, endangered species or rare habitat. This approach involves strategies such as maintaining so-called stepping stones of habitat that enable wildlife to migrate through hostile landscapes.

Corridors are a landscape element that plays a key role in connectivity. Narrowly defined, corridors are strips of habitat that differ from the adjacent habitat on both sides (Forman 1995) or, alternatively, that serve as linear linkages between larger habitat blocks (Forman and Godron 1986). Also included in the corridor concept are artificial structures such as tunnels or underpasses that are designed to permit animal movement (Noss 1993). A broader way to define corridors is according to their purported functions, such as facilitating the flow of nutrients across the landscape (Forman 1995) as well as providing routes for movement or gene flow between separated populations (Harris and Scheck 1991) or even for the movement of entire biological communities (Hobbs and Hopkins 1991). Linked to this latter idea, scientists have evoked corridors as a potentially promising tool to enable biological communities to respond to habitat and climate change (Harris and Gallagher 1989). Lastly, scientists may define corridors according to their conservation status. Corridors fall into four general conservation types (Harris and Gallagher 1989, Loney and Hobbs 1991, Bennett 1999): (i) natural corridors, such as waterways and streams and their associated riparian vegetation; (ii) remnant corridors, such as strips of unlogged forest within clearcuts, natural woodlands along roadsides, and natural habitats retained as links between nature reserves; (iii) regenerated corridors, such as fencerows and hedges; and (iv) planted corridors, such as windbreaks or shelterbreaks and urban greenways.

In short, while generally acknowledging their important connective functions, scientists do not always agree on what is, or is not, a corridor.

In the meantime, however, corridors have emerged as a widespread strategy for countering fragmentation, sparking the imagination of conservation practitioners and a wider public concerned with biodiversity loss. Corridors are appealing because they purportedly maintain or restore the very connectivity that fragmentation undermines. Whether in a backyard or across a landscape, everyone can visualize a corridor and associate it with connectivity. This probably contributes to the growing popular support for the concept of corridors in many countries.

Ongoing corridor projects worldwide range in scale from small revegetation projects led by local communities to ambitious continent-wide schemes that aim to redesign the way people use and manage the land. Some projects involve narrow tunnels and overpasses that help animals move across local barriers such as roads and railroads. Others include patchworks of whole ecosystems, linking landscapes to conserve wide-ranging species and critical ecological processes.

For practical purposes, we can simply define corridors here as *spaces in which connectivity between species, ecosystems, and ecological processes is maintained or restored at various scales.* This broad definition incorporates the full range of functions, conditions, and scales variously attributed to corridors in the scientific literature, and it reflects widespread practices in conservation. Because the lay public readily understands the term "corridor" as a connecting space, expanding the term to include diverse conservation initiatives promoting connectivity at different scales offers an appealing way to include a variety of objectives and to attract broad support. For clarity in referring to corridors with specific configurations, functions, and scales, we can apply descriptors to specific types of corridors and corridor initiatives.

The broad definition of corridors provided above reflects an ongoing trend to plan and implement biodiversity conservation at increasingly larger scales. Until the 1970s, conservation efforts focused mostly on establishing and maintaining discrete protected areas. The 1982 World Congress on National Parks in Bali endorsed a broader approach that encompassed adjacent populations and buffer zones (McNeely and Miller 1984). This then led during the 1980s to the widespread application of integrated conservation and development projects. Unfortunately, these projects generated mixed results and usually failed to address the root causes of biodiversity loss (Brandon et al. 1998, Wells et al. 1999). Furthermore, even though protected areas cover approximately 11.5 percent of the land surface, scientists increasingly agree that they are insufficient to protect the full range of biodiversity and its associated processes (Chape et al. 2003). These considerations — combined with the urgent need to establish priorities for biodiversity conservation based on solid scientific principles — led to still larger-scale approaches such as ecoregion conservation.[4]

As a practical, on-the-ground strategy for restoring landscape connectivity, corridors provide an important tool for implementing these large-scale approaches to conservation. They are more than just linear protected areas. Corridors link natural and seminatural habitats through landscapes dominated by human activities. In some cases, the corridors themselves may largely be unnatural, with their main function being to enhance other, more natural areas. Perhaps most significantly, corridors require conservation practitioners to look beyond core protected areas and address a wide range of issues affecting resource-use decisions by people.

In short, corridors are both an integral part of the evolution of conservation biology to larger scales and highly complementary to such approaches. While large-scale conservation provides an encompassing biodiversity vision that defines ambitious goals, corridors furnish an operational means to achieve those goals.

Conservation biologists and land-use managers are planning hundreds of corridor initiatives worldwide. However, there is relatively little technical information to guide the design and implementation of these initiatives. The acceptance and application of corridors as a strategy for biodiversity conservation has largely outpaced the collection of empirical data to support and manage them. This includes data on the requirements of target species, communities, and ecosystems and how corridors might benefit them. While the scientific literature on corridors has mushroomed in recent years, there are still critical information gaps. This is especially true at large scales, where corridors potentially may be most beneficial. Furthermore, there is almost no published information on the socioeconomic and political issues related to corridors. Such information would provide essential guidance for design and implementation in a context of competing interests and trade-offs.

Contents of the Book

This book focuses on corridors designed to contribute, either directly or indirectly, to biodiversity conservation. This may be through maintenance or restoration either of biodiversity per se or of the ecological functions on which biodiversity depends. An exclusive focus on biodiversity conservation, however, ignores opportunities for generating other benefits, such as increased agricultural productivity through soil restoration, improved drinking water, and recreation. Including these other objectives is important for building broad-based support for corridors, yet it will have implications for their design and implementation. These implications may not complement biodiversity conservation and in some cases may run counter to it.

This book provides an overview of current knowledge on corridors, their design, and their implementation, based on information obtained from diverse sources. Chapter 2 examines the corridor concept, associated corridor terms and typologies, and the scientific debate over corridors. That debate has focused on the evidence for the purported functions of corridors, their positive and negative effects, and their cost-effectiveness. This chapter derives from a selective review of the scientific literature and provides a foundation for the rest of the book.

Two major types of corridors are discussed here:

Linear corridors, which

- provide a relatively straight linkage between two or more larger habitat blocks, typically over distances of meters to tens of kilometers,

- are designed to maintain or restore target species, movement of short-ranged animals, and/or local ecosystem services,
- are most relevant in relatively disturbed landscapes, and
- are established using relatively straightforward strategies such as purchasing land or easements and/or strictly enforced zoning;

and *Landscape corridors,* which

- provide multidirectional connections between a mosaic of ecosystems that cover areas from one to thousands of square kilometers,
- are designed to maintain or restore entire biota, movement of far-ranging species, a full range of landscape mosaics, and/or ecosystem services at a regional scale,
- are most relevant in relatively intact landscapes (but can be usefully applied regardless of landscape condition), and
- require a suite of approaches such as land-use zoning, establishment of public and private protected areas, application of incentives for environmentally benign resource and habitat use, and purchase of land or easements.

Landscape corridors provide an appropriate scale and context for planning connectivity over the long term, while linear corridors serve as the building blocks for expanding or maintaining that connectivity over shorter scales of space and time. Utilizing both allows practitioners to strike a balance between more general, long-term goals needed for planning and more localized, short-term objectives that can motivate stakeholders.

Scientific debate over corridors has focused on three main points:

- *Positive and negative effects.* An array of benefits has been attributed to corridors, ranging from conservation of biodiversity and ecological processes to support of agriculture, forestry, recreation, and aesthetics. Yet corridors may also serve as pathways for fire, predators, and pathogens, which can undermine conservation objectives. On balance, most scientists concur that the potential benefits of corridors do outweigh the costs.
- *Cost-effectiveness.* Especially when they involve habitat restoration, corridors can be a costly enterprise, which has led some critics to question their cost-effectiveness. Cost–benefit analyses, however, should incorporate the full range of benefits that corridors provide, and consider the costs of alternatives, such as translocation of wildlife.
- *Evidence of corridor functions.* Most research on corridor functions to date has focused on small-scale processes such as animal movement over

short distances. This is due to the formidable difficulty of carrying out experimental studies at larger scales of space and time. Although there is little empirical evidence demonstrating that corridors function at such scales, the weight of existing evidence shows that isolation of populations and communities through loss of intervening habitat is detrimental.

Chapter 3 examines corridor design, using theoretical considerations, empirical evidence derived from the literature, and practical experiences gleaned from cases worldwide. It is intended to provide insights for scientists and resource managers who must address the challenges of designing corridors. Corridor design is highly specific to the unique requirements of the species, habitats, ecosystems, and ecological processes of concern in each case. Design must address not only biophysical elements but also socioeconomic and political factors. Designing corridors only makes sense within a larger context, such as ecoregion conservation. Ideally, a biodiversity vision should be developed *before* a corridor is designed in order to allow scarce resources to be focused on key biodiversity targets and priority areas.

Defining explicit corridor objectives is also imperative for design. Where possible, corridor design should work to achieve various objectives at various scales, although in practice multiple uses sometimes conflict. Protecting key species is the most frequent objective of biological corridors and often determines their design. The life histories of focal species — particularly homeranges and dispersal patterns — have critical implications for corridor design. By targeting top carnivores or other so-called keystone species, a focal species approach assumes that numerous other species will also gain protection. Corridor design should also achieve broader objectives by including maximum variability of such features as biotic communities, ecosystems, soil types, topographic gradients, and special landscape elements. Although largely untested, properly designed corridors also present one of the few tools available to mitigate the impacts of climate change on biotic communities.

The book breaks entirely new ground in chapter 4, which explores key factors that are likely to influence corridor implementation, once again drawing on actual cases. Five questions are of special interest here:

• *How do you manage corridors?* Management of corridors depends largely on their design, which in turn reflects their primary conservation objectives. Corridor design may focus on facilitating the movement of endangered species or, alternatively, on addressing large-scale processes such as climate change. Achieving these objectives requires progressively more complex management.

- *What are the obstacles to implementing corridors?* Here the book examines diverse obstacles to corridor implementation, including the drivers of biodiversity loss, lack of public awareness, resistance to decreased resource control and use, economic costs, and the high uncertainty associated with corridors.
- *How do you build support for corridors?* The challenge here is to involve a wide range of frequently conflicting interest groups — including resource owners or users, business sectors, policymakers, and other potential allies or opponents — to ensure the necessary buy-in for implementing corridors.
- *What are the incentives for implementing corridors?* To overcome a major obstacle to corridor implementation, incentives frequently are necessary to compensate resource owners or users for the income or other benefits lost. Negative incentives may involve the potential loss of critical ecosystem services and treasured species and habitats. Such possibilities can provide powerful motivation for corridor initiatives. Positive incentives involve the actual benefits that corridor implementation can provide, and include private revenue from land sales, swaps, or easements; tax breaks; tourism; increased government revenue; and maintenance or restoration of critical environmental services.
- *How should corridors be governed?*[5] Especially at large scales, corridors are ambitious undertakings and need support from a wide range of interest groups. Governing a corridor requires defining the roles and responsibilities of the diverse interest groups (both public and private) involved in its planning and implementation.

Finally, chapter 5 provides case studies of actual corridor initiatives worldwide. Because there is virtually no analysis in the literature of the socioeconomic issues related to corridors, these cases provided a supplementary source of information for corridor design (chapter 3) and served as the primary source of information on corridor implementation (chapter 4). The cases are:

Case 1: Atlantic Forest corridor, Brazil
Case 2: Talamanca corridor, Costa Rica
Case 3: Pinhook and Panther Glade corridors, Florida, USA
Case 4: Yellowstone to Yukon (Y2Y) corridor, USA and Canada
Case 5: Klamath–Siskiyou corridor, USA
Case 6: Lower Kinabatangan River corridor, Sabah, Malaysia
Case 7: Terai Arc corridor, India and Nepal
Case 8: Veluwe corridor, the Netherlands

The Book in Context

This book is not the first to provide a synthesis of knowledge on corridors. Several scientific symposia and reviews have addressed this theme (e.g., Mackintosh 1989, Little 1990, Hudson 1991, Saunders and Hobbs 1991a, Smith and Hellmund 1993). In particular, a review by Andrew Bennett (1999), entitled *Linkages in the Landscape: The Role of Corridors and Connectivity in Wildlife Conservation*, provides an extremely useful source for the scientific and technical aspects of corridors as well as numerous examples of corridor projects worldwide.

None of these references, however, attempts to complement coverage of scientific and technical aspects of corridors with an examination of their attendant socioeconomic and political issues. It is precisely these issues that typically require the most time and resources for corridor design and implementation, and they will probably have the greatest impact on corridor initiatives over the long term.

This book aims to strike a balance between the scientific and social aspects of corridor establishment and implementation. To achieve such a balance, we summarize the scientific aspects and present them in a form that should be comprehensible to nonscientists. Corridors are the subject of a huge and rapidly growing body of literature in conservation biology. Readers wishing to access that literature should seek the references cited above as well as recent issues of *Conservation Biology* and similar scientific journals. By the same token, the book provides an overview of socioeconomic and political issues that should be accessible to biologists and other scientists.

By thus providing a balanced and relatively condensed coverage of the key issues involving corridors, the book targets a broad audience of conservation practitioners, decision makers, and laypersons interested in conservation issues in general and corridors in particular. In addition to providing a multifaceted analysis of corridors, the book attempts to make the analysis useful by combining a succinct review of the conceptual issues with tangible examples of how corridor design and implementation play out in the real world.

2 Conceptual Foundations of Corridors

Corridors are bandages for a wounded natural landscape.
— Soulé and Gilpin (1991:8)

This chapter reviews the foundations of the corridor concept, examines the wide array of terms associated with corridors and suggests their division into two basic types, and then assesses the scientific debate regarding corridor functions, benefits, and costs.

Conceptual Foundations

The concept of corridor implementation as a conservation strategy derives from the assumption that maintaining or restoring connectivity at diverse scales is essential for conserving biodiversity in increasingly fragmented natural ecosystems and communities. The significance of connectivity for conservation rests on three conceptual frameworks: the equilibrium theory of island biogeography, the dynamics of populations separated into habitat patches ("metapopulations"), and principles of landscape ecology. In the sections below, we examine each of these frameworks and its significance in explaining species assemblages in habitat fragments, and the potential roles of connectivity in maintaining or restoring those assemblages.

The Equilibrium Theory of Island Biogeography

Wilson and Willis (1975) proposed the corridor concept based on the equilibrium theory of island biogeography (MacArthur and Wilson 1967).

Observations showed that larger islands, and islands located closer to a main-land, supported higher levels of species richness than did smaller, more isolated islands. The equilibrium theory explained these patterns in terms of immigration (or colonization) and extinction: an island's degree of iso-lation determines the rate of immigration, its area determines the rate of extinction, and its number of species represents a balance (or equilibrium) between these two processes.

Biologists reasoned that the equilibrium theory might also be applied to terrestrial fragments or patches[1] of habitat, which were often described as hospitable "islands" surrounded by a "sea," or matrix, of inhospitable land-scape. Based on the equilibrium theory, a newly isolated habitat patch would have more species than it could maintain. It would initially sponsor most of the species from the original unfragmented habitat. However, because of the smaller amount of habitat in the isolated patch, the individual species would have smaller populations relative to their original total population in the unfragmented landscape. These smaller populations would be more vul-nerable to extinction, and this would lead to an overall increase in the local extinction rate. As long as immigration remained constant, the higher ex-tinction rate would eventually push the total number of species to a new, lower equilibrium.

Scientists considered corridors to be one way to partially counteract this higher extinction rate. By reducing the physical barriers to immigration, a corridor would shorten the expected time for a species to colonize or reco-lonize a habitat patch where that species had gone extinct. This would ef-fectively raise the equilibrium number of species. The increased immigra-tion might also supplement small local populations and prevent their extinction in the first place. For a more detailed discussion of these early ideas, see Diamond 1975.

Since it was proposed, however, scientists no longer find the equilibrium theory of island biogeography adequate for explaining terrestrial fragmen-tation effects, because habitat patches differ from true islands in how they interact with their surroundings. For example, in contrast to islands, habitat patches (and their resident species) are more susceptible to processes origi-nating from the surrounding matrix, such as fire and species invasions (e.g., Gascon and Lovejoy 1998). Such processes tend to confound area and de-gree of isolation as causal factors in explaining species assemblages in non-insular habitat patches. The dynamic nature of fragmentation, and the widely varying response rates of different species to it, are other confounding factors.[2] As a result, today most conservation biologists believe that the equilibrium theory of island biogeography alone cannot account adequately for the effects of fragmentation or for the potential roles of corridors in diminishing those effects (Simberloff and Abele 1982, Zimmerman and Bierregaard 1986, Bierregaard et al. 1992).

Metapopulation Theory

Another conceptual framework important to the corridor concept is that of *metapopulations* — a term for populations that are subdivided among separated habitat patches but that interact with each other (Levins 1969, 1970; Hanski 1989, 1999; Hanski and Gilpin 1991, 1997; McCullough 1996). In this species-level approach to conservation in fragmented landscapes, species movements supplement local populations in decline, recolonize habitats where local populations have disappeared, and colonize new habitats as they become available (Bennett 1999).

Metapopulation theory is a rapidly expanding area of research and one that conservation scientists increasingly call upon, especially in cases where human activities have extensively fragmented a species' range. Under such conditions, a formerly continuously distributed regional population becomes limited to a series of shrinking habitat patches, each containing a smaller local population. These smaller populations are intrinsically more vulnerable to extinction because it is easier for disasters such as fires or hurricanes to completely eliminate them. Even simple variations in population parameters, such as when one sex disappears through demographic "accidents," can become problematic for very small populations. Metapopulation theory posits that movement of organisms among patches increases the stability of a regional population. This is because movement allows interbreeding between local populations and immigration to fragments where a species is declining or has become extinct. This movement of individuals would reduce the ultimate effect of some of the threats described above. Consequently, habitat configurations that assist movements of organisms through the landscape should contribute to the persistence of species. By focusing on habitat configurations, the metapopulation concept thus offered the possibility of explaining species' population dynamics and, ultimately, the factors that contribute to their persistence in a landscape.

While conservationists often invoke metapopulation theory as an explanatory framework in the conservation literature, there is relatively little empirical evidence supporting it (Simberloff and Cox 1987, Simberloff et al. 1992). The extinction and colonization events that make a population an operating "metapopulation" are typically very difficult to observe (Mann and Plummer 1995). For many species, these events may be so rare that biologists are unlikely ever to witness them in the field. Furthermore, observations show that the persistence of local populations depends not only on habitat configurations but on factors such as the rate at which those configurations change, the quality of habitat patches, and the dispersal capacity of species (e.g., Hanski 1989, Fahrig and Merriam 1994).

Because of the difficulty of studying natural metapopulations, much of the research has focused on theoretical models and microcosms (e.g., Kareiva 1990, Fahrig and Merriam 1994, Burkey 1997, Gonzalez et al. 1998). Of the empirical evidence available, some of the best comes from such extensively studied butterflies as the Glanville fritillary (*Melitaea cinxia*) and Edith's checkerspot (*Euphydryas editha*) (Ehrlich and Hanski 2004), as well as some studies of spotted owls (*Strix occidentalis*) (Lahaye et al. 1994, Akçakaya and Raphael 1998). Unfortunately, for only very few species is there the opportunity for the intensive, long-term studies necessary for understanding a metapopulation and the potential effects of corridors. What evidence there is does support the metapopulation theory, but there is still uncertainty on how common metapopulations are in nature. In short, metapopulation theory has much appeal as a theoretical conservation tool, but faces obstacles in its application to conservation of most species.

Landscape Ecology

A third and more practical framework to explain the role of connectivity in fragmented landscapes is the field of *landscape ecology*. Landscape ecology moves beyond species to focus on entire landscapes and how their structure influences species and ecosystem processes (Forman and Godron 1986, Forman 1995, Turner et al. 2001). It explicitly recognizes that a landscape is an integrated system, and that you cannot fully understand one piece if you ignore the other pieces to which it is connected.

Whether pristine or heavily modified, all landscapes are mosaics comprised of three basic structural elements, illustrated schematically in figure 2.1:

- *patches* or *fragments* are habitat types embedded in the more widespread matrix;

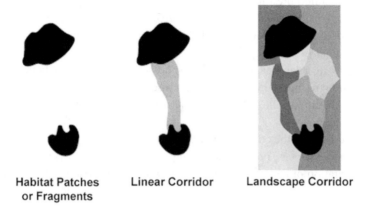

Habitat Patches   Linear Corridor   Landscape Corridor
or Fragments

FIGURE 2.1

The elements comprising the two main corridor types. A linear corridor (center) consists of a narrow, unidirectional linkage between two habitat patches or fragments, whereas a landscape corridor (right) consists of multidirectional linkages within a matrix comprised of multiple habitat types. Three basic landscape elements — patches or fragments, corridors, and the matrix — together comprise a landscape mosaic. Courtesy of Conservation International (2000).

- *corridors* connect original habitats and/or habitat fragments;
- the *matrix* is the most widespread habitat within which other elements are embedded, and it can be either the original habitat type (e.g., a primary forest now surrounded by a matrix of agriculture) or a modified one (e.g., an agricultural plantation surrounded by a matrix of pristine forest).

The movement of organisms, materials, and energy through these structural elements is a critical landscape function. This *conduit* function is influenced, in turn, by the structure of the landscape. The most direct movements are in linear habitats such as rivers and streams. They provide a downstream conduit for energy, nutrients, and organisms such as aquatic invertebrates and plant propagules, while some migratory fish and mammals can swim long distances upstream. Nonlinear features of the landscape can also influence the conduit function. For example, some rivers flood annually and distribute nutrients across wide floodplains. The damming of such a river could interrupt not only its linear conduit function, but also the wide-ranging flow of nutrients across the landscape.

Scientists envision corridors as key strategies for restoring or enhancing the conduit function in altered landscapes. For example, linking remnants of the Atlantic Forest in Brazil could permit interbreeding of currently isolated populations of the highly endangered golden lion tamarin (*Leontopithecus rosalia*). This may be essential for the long-term future of this species, whose small isolated populations face a risk of inbreeding. However, many processes in a landscape intertwine. The same corridor that benefits the tamarin by reducing inbreeding might also serve as a conduit for wildfires that would threaten the little habitat that remains for this species (see chapter 5, case 1).

In addition to the conduit function, other landscape functions performed by corridors are best understood through landscape ecology (see fig. 2.2 for an illustration of these functions as exemplified by a stream system). A corridor can serve:

• As *habitat*: by providing appropriate environmental conditions for organisms to live. For example, the Cascade–Siskiyou National Monument in the Klamath–Siskiyou ecoregion provides habitat for many local species while also providing a critical corridor between two mountain ranges with habitat for bears, cougars, and other species (case 5).
• As *barrier* or *filter*: by being a partial or complete obstacle to the movement of organisms and abiotic materials, a corridor may serve as a *barrier*. In many cases, sharp transitions between distinctive habitats such as forests

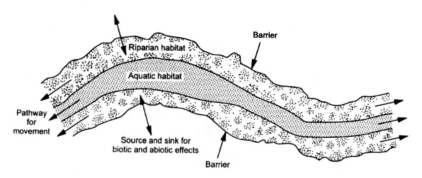

FIGURE 2.2

Landscape functions performed by corridors. This schematic representation of a stream system illustrates many of the functions performed by corridors. The riparian and aquatic zones: (i) provide a habitat for organisms, (ii) serve as a pathway or conduit for movement of organisms and abiotic materials, (iii) pose a barrier or a filter to the movement of organisms and abiotic materials, and (iv) act as a source or a sink for exchanges of organisms and abiotic materials with the surrounding environment. Courtesy of Bennett (1999).

and savannas, or rivers and arid environments, obstruct movement. This "barrier" function can act to keep animals targeted for conservation within a corridor while also protecting corridor habitats from invasive species in the matrix. In a similar manner, a corridor of small habitat fragments between larger habitat blocks, a so-called stepping stone corridor, can act as a selective *filter* for species. It may allow movements by disturbance-tolerant animals yet fail to permit movement by less tolerant species. For example, remnant riverine forests of the Lower Kinabatangan River in Malaysia permit movement of only part of the original fauna (case 6).

• As *source* or *sink*: by providing a source from which organisms and abiotic materials spread to surrounding areas, or by existing as a sink that is dependent on organisms and abiotic materials arriving from elsewhere (Pulliam 1988). For example, forested areas with relatively large and growing populations of golden lion tamarins provide a *source* for translocation to areas without resident populations. Conversely, small or low-quality forest fragments may function as *sinks* for golden lion tamarins if they are unable to maintain a population without emigration from a source habitat (both examples from case 1). For conservation, it is important to identify the source habitats as priority targets for conservation actions.

Each of these landscape-level corridor functions may or may not be desirable. For example, the conduit function can be positive when it promotes the movement of key species, but negative when it contributes to the mortality of those species by providing a conduit for fires or disease (Simberloff and Cox 1987). Corridors that retain excess nutrients and sediment can serve a beneficial sink function for aquatic ecosystems (Binford and Buchenau 1993). If a corridor is too narrow, however, it can become a mortality sink for wildlife by increasing their vulnerability to hunting or predation (Simberloff and Cox 1987, Thorne 1993). In short, the functions that a corridor fulfills depend on its interaction with surrounding matrix and patch environments, as well as its internal dynamics.

By including both habitat fragments and their surrounding matrix, and not only populations but the critical ecological processes that link them, landscape ecology provides a comprehensive framework for understanding the effects of fragmentation and the potential role of corridors in reversing those effects. Even though the origins of the corridor concept are with island biogeography, and somewhat with metapopulation theory, landscape ecology provides a more practical conceptual framework for the understanding and application of corridors.

Terminology and Types of Corridors

The scientific and popular literature uses a perplexing array of descriptors in reference to corridors and corridor initiatives. Based on these descriptors, we define five broad yet overlapping groups of corridors:

• A "habitat corridor" is a linear strip of native habitat linking two larger blocks of the same habitat. Such corridors are primarily for enhancing the protection of, and increasing the area of, rare habitats. They often involve linkages between small-scale protected areas or local restoration projects linking two separated blocks of habitat.
• The term "ecological corridor" refers to corridors designed primarily to maintain or restore ecological services upon which biodiversity conservation depends, such as maintenance or restoration of soil or water quality. Other services, such as recreation or aesthetics, may be a secondary objective.[3] The term "ecological corridor" is sometimes employed synonymously with "biological corridor" or "biodiversity corridor" (see below).
• "Movement corridor," "dispersal corridor," and "wildlife corridor" refer to corridors designed primarily to promote the movements or migrations of specific species or groups of species. Depending on the species involved, the dimensions of such corridors could range from linear connections of less than one hundred meters to large swaths of landscape covering hundreds or even thousands of square kilometers. In some cases, they consist of ingenious artificial connections such as bridges or tunnels specifically designed to facilitate animal movements over or under roads and other obstacles. The primary axis of such corridors corresponds to the main direction of the species' movements or migrations.
• "Corridor networks" are a system of corridors running in multiple directions. The term usually applies to linkages established in hostile and/or highly degraded landscapes. Scores of such projects are currently under way in Europe (Bennett and Wit 2001).
• "Biological corridor," "biodiversity corridor," and "ecological corridor" refer to large-scale landscape linkages covering hundreds to thousands of square kilometers. This includes mega-initiatives such as the Mesoamerican Biological Corridor (Miller et al. 2001).

These terms span the full range of corridor objectives related to biodiversity conservation — including maintenance or restoration of habitat linkages, ecological services, and species' movements or migrations.

The considerable overlap in the use of corridor descriptors reflects their multiple purposes, configurations, and spatial scales. An alternative and simpler scheme, based on a landscape perspective, distinguishes two basic types of corridors (fig. 2.1):

• *Linear corridors* establish or maintain relatively straight-line connections between larger habitat blocks and extend over distances of up to tens of kilometers. A seven-mile corridor linking the Loxahatchee National Wildlife Refuge and J. W. Corbett Wildlife Management Area in southeast Florida illustrates this approach (Harris and Gallagher 1989). Linear corridors are appropriate for accomplishing a few specific objectives, such as facilitating movement of target species. They are likely to be effective in situations where:

>   a large part of the landscape is modified and inhospitable to native
>   species,
>   the species of concern are dependent on relatively undisturbed habi-
>   tats, and/or
>   the desired habitat or ecosystem can be maintained or restored in
>   linear habitats such as fencerows or streams.

• *Landscape corridors* maintain or establish multidirectional connections over entire landscapes and can encompass up to thousands of square kilometers. A landscape corridor may contain *multiple* linear corridors. It is in essence a regional planning unit designed to maximize the connectivity of habitats, ecosystems, and ecological processes at large scales (Conservation International 2000). The Mesoamerican Biological Corridor — which aims to promote ecological connectivity at a subcontinental scale by establishing a mosaic of natural habitats and compatible resource uses — is a classic example of this approach. The design of landscape corridors may address a broad array of objectives involving biodiversity, ecological services, resource uses, recreation and aesthetics, and climate change. From a biodiversity perspective, landscape corridors are likely to be effective in situations where:

>   a significant portion of the landscape is intact (although corridors still
>   provide a useful framework in altered landscapes),
>   target species require large areas of habitat, and/or
>   the species or biotic communities of concern have a high tolerance to
>   the existing landscape matrix.

In short, landscape corridors provide an appropriate scale and context for planning connectivity, while linear corridors serve more as operational building blocks for restoring that connectivity.

In a similar way, corridors are complementary to large-scale approaches such as ecoregion conservation. Ecoregion conservation defines an encompassing biodiversity vision that defines critical targets.[4] Corridors provide one operational means to attain some of those targets. Nevertheless, ecoregions and corridors play distinct roles in conservation. Ecoregions are distinguishable based on biogeographic criteria such as floristic composition or topography, whereas corridors are defined in terms of their role in enhancing habitat connectivity. Ecoregions provide a useful conceptual framework and scale for identifying priorities for biodiversity conservation, defining appropriate interventions, and monitoring their impact. In contrast, corridors represent one kind of conservation intervention. In practice, corridors are generally smaller than ecoregions, although in some cases — such as the Yellowstone to Yukon (Y2Y) Corridor and the Mesoamerican Biological Corridor — landscape corridors may exceed ecoregions in extent.

Drawing from discussions above and in subsequent sections of this book, box 2.1 summarizes the definitions, objectives, and approaches to implementation of linear and landscape corridors and gives examples of each.

BOX 2.1. CHARACTERISTICS OF THE TWO BASIC CORRIDOR TYPES

Linear (or Habitat, or Movement) Corridors

- *Definition*: provide a single, continuous (or near-continuous) link between two or more usually larger habitat blocks, typically over distances of up to tens of kilometers.
- *Objectives*: maintain or restore target species, movement of short-ranged animals, linkage of habitat fragments, and/or local ecosystem services.
- *Instruments*: use specific strategies such as purchasing land or easements and/ or strictly enforced zoning, which are likely to be more viable in restricted areas.
- *Examples from Cases*:

   Ecoducts in Veluwe, the Netherlands (see fig. 4.1 and case 8)
   Lower Kinabatangan River elephant corridor in Sabah, Malaysia (case 6)
   Atlantic Forest corridors for golden and black lion tamarins in Brazil (case 1 and box 4.4)
   Panthers in south Florida, USA (case 3)
   Bow Valley Corridor for bears and elk in Alberta, Canada (case 4)

continued

Landscape Corridors

- *Definition:* provide multidirectional connections between a mosaic of ecosystems that extend over an area of tens to thousands of square kilometers.
- *Objectives:* maintain or restore entire biota, movement of far-ranging species, linkage of habitat or ecosystem mosaics, and/or ecosystem services at a wide scale.
- *Instruments:* require regional planning with land-use zoning, establishment of public and private protected areas, and application of incentives for environmentally benign resource and habitat use. Purchases of land or easements, or designation of strict protected areas, restricted to areas critical for biodiversity conservation.
- *Examples from Cases:*

  Corridor network in and beyond Veluwe, the Netherlands (case 8)
  Restoration of tiger home-ranges in Terai Arc in India and Nepal (case 7)
  Yellowstone to Yukon Corridor in Canada and the United States (case 4)
  Cascade–Siskiyou National Monument (case 5)
  Talamanca corridor, Costa Rica (case 2)

Corridor Controversy

Much debate has been generated among conservation biologists regarding the functions, advantages and disadvantages, and cost-effectiveness of corridors in relation to other conservation strategies. Both proponents (e.g., Noss 1987, Harris and Gallagher 1989, Beier and Noss 1998) and critics (e.g., Simberloff and Cox 1987, Simberloff et al. 1992, Hobbs 1992, Mann and Plummer 1993 and 1995) acknowledge the enormous threat posed by habitat fragmentation. However, they disagree about whether corridors represent the best, or even an effective way, to address this threat. The debate has focused on three main points:

the scientific evidence for corridor functions,
the positive and negative effects of corridors, and
the cost-effectiveness of corridors.

Scientific Evidence for Corridor Functions

Proponents have attributed an array of benefits to corridors at both local and landscape levels (box 2.2). The diverse values provided by landscape connectivity for conservation of biological diversity, land uses, and aesthetics are indisputable. Where scientists have disagreed is whether corridors — and in particular relatively small-scale, linear corridors — are the best or even a sound strategy for mitigating the effects of habitat fragmentation (Harrison and Bruna 1999).

The debate has focused on the capacity of linear corridors to facilitate animal movement in fragmented landscapes, and the potential role of such corridors in conserving animal populations and their habitat. Much of the early debate on corridors focused on either theoretical considerations or empirical observations showing (or failing to show) animal movement through habitat corridors (Noss 1987, Simberloff and Cox 1987, Hobbs 1992, Simberloff et al. 1992). With growing interest in the potential conservation values of corridors (probably sparked by this debate), scientists have carried out more research on corridor functions in recent years. Yet a review of thirty-two such studies revealed that only twelve allow meaningful inferences about corridor functions. Only ten of those offer persuasive evidence of increased movement or dispersal (Beier and Noss 1998).

BOX 2.2. BENEFITS OF LANDSCAPE CONNECTIVITY

Biological Diversity

Maintenance or increase of species richness and diversity
Recolonization following local species extinctions
Colonization of new or recovered habitat
Enhancement of gene flow to minimize inbreeding depression in small
    populations
Pathway for redistribution of entire communities, as during seasonal migra-
    tions or in response to climate change

Agriculture and Forestry

Windbreaks for crops, pasture, and livestock
Reduction of soil erosion by wind and water
Source of forest products (timber, firewood, fruits, seeds, latex, etc.)

Water Resources

Maintenance of groundwater levels and quality
Flood regulation and mitigation

Recreation and Aesthetics

Wildlife observation and environmental education
Hiking and camping

— Adapted from Forman 1995 and Bennett 1999

A major problem of many empirical studies is that they focus exclusively on the behavioral and population responses of animals within corridors. Such studies therefore fail to provide comparative information from the surrounding matrix (Simberloff et al. 1992, Beier and Noss 1998). In addition, while research on corridors is growing rapidly, most studies have focused on small-sized animals such as rodents and on local-scale movements. This approach reflects the formidable difficulties of carrying out experimental studies at larger scales. With few exceptions (e.g., Beier 1993), corridor studies have not involved larger and wider-ranging species, such as top carnivores, which tend to play critical roles in biological communities. While local-scale movements of small animals may shed light on corridor functions, the key challenges facing conservation today are at landscape or regional scales. It is at such scales that corridors potentially can provide maximum benefits.

Research to date indicates that animal responses to corridors are highly species- and scale-specific. Habitat generalists such as white-tailed deer (*Odocoileus virginianus*) are capable of exploiting large portions of a landscape and move through both narrow habitat corridors and the disturbed, surrounding matrix. In contrast, understory birds tend to be far less habitat-tolerant. They may use corridors linking similar forested habitats only under extreme conditions. The response of any given species also seems to vary according to the scale under consideration. As a result, designing research to address general questions about corridor viability — such as "principles that predict behavioral and population responses to corridors across species and landscapes" (Haddad et al. 2000) — is probably not realistic. Instead, corridor research needs to focus on the details of particular species and scales relevant to conservation practice (Noss and Beier 2000).

Solid experimental research on the values of corridors is still lacking. As mentioned above, virtually all experimental research has focused on behavioral and population responses of specific animals. Scientists have not explored the community- or ecosystem-level effects of corridors on such factors as disturbance, exotic species invasions, predation, and species richness (Beier and Noss 1998). Lack of experimental evidence has led some critics to claim that scientists risk losing credibility with policymakers if they advocate corridors (Mann and Plummer 1995). The reverse of this critique may also be valid. Rather than making a case against corridors, the scarcity of scientific data in fact justifies more and better research (Hobbs 1992).

In the meantime, the following factors merit consideration (Bennett 1999):

- Currently there is no evidence that corridors have negative impacts on plant or animal populations, communities, or ecosystems (Beier and Noss 1998).

- In the absence of complete information, it is safer to assume that the natural condition (a connected landscape) is preferable to an artificial condition (a fragmented landscape) (Saunders and Hobbs 1991a, Harris and Scheck 1991, Beier and Noss 1998).
- The weight of existing evidence shows that isolation of populations and communities, through loss of intervening habitat, is detrimental to biodiversity (Noss 1987, Harris and Gallagher 1989).
- The rapidity of habitat fragmentation does not give us the luxury of waiting to take action until we acquire complete evidence (e.g., Harris and Scheck 1991).

Positive and Negative Effects of Corridors

Proponents and critics have disagreed about whether at the population level corridors can increase effective population size, thereby reducing extinction risk due to factors such as inbreeding depression (e.g., Harris and Gallagher 1989, Noss 1987, Simberloff and Cox 1987, Simberloff et al. 1992). Simberloff and Cox (1987) suggest that in some cases translocation may be more effective for conserving the genetic diversity of small populations (box 2.3).

BOX 2.3. CONTRASTING STRATEGIES FOR CONSERVING
THE FLORIDA PANTHER

The Florida panther (*Felis concolor coryi*) is one of the most endangered species in North America. Because of rapid development, its range is now limited to a few isolated locales (including the one million–hectare Everglades/Big Cypress protected areas) that support tiny populations at high risk of inbreeding depression. To counter this threat, efforts to establish corridors to connect remaining habitat fragments are now under way (see case 3).

Conservation biologists, however, disagree about the efficacy of this approach. Simberloff and Cox (1987) argue that the most effective, and least costly strategy, would be to translocate individuals between reserves as a way to mitigate inbreeding depression. Translocation, however, is a complex and frequently risky strategy for conservation (Griffith et al. 1989). It is also not a viable strategy for uniting multiple species in complex systems. Noss (1987) believes that the life history traits of the Florida panther — in particular its sensitivity to fragmentation and its need for immense areas of intact habitat — indicate that increased landscape connectivity may be the best if not the only way to preserve this species.

Regarding their effects on ecological processes, Noss (1987) notes that corridors can provide escape routes from predators and fire. In contrast, Simberloff and Cox (1987) observe that corridors can facilitate the spread of fire, disease, and pests, and that they enhance vulnerability to predators and other mortality threats. To date, however, there is little or no evidence indicating increased mortality due to the spread of disease or parasitism through corridors,[5] although they do act as conduits for fire (Bennett 1999; see case 1).

Proponents also argue that corridors will increasingly be necessary to allow movement of entire biotic communities in response to climate change (Harris and Gallagher 1989, Hobbs and Hopkins 1991, Noss 1991). To do this, however, corridors would have to be designed to include space for each species in the community to live and breed as well as to move (Simberloff et al. 1992). Because each species has distinct ecological requirements, in practice corridors designed for climate change probably would act as a selective filter, allowing some but not all species to migrate (Hobbs and Hopkins 1991). This might be especially true in cases where environmental gradients flow in differing directions. In southwestern Australia, for example, the different directions of environmental gradients might mean that some species would shift their distributions in different directions, depending on which environmental factor is most important for them (see fig. 3.1). This could mean that different species, or groups of species, would need separate corridors to shift to their new locations.

Cost-Effectiveness of Corridors

Critics argue that corridors tend to be ambitious undertakings that are likely to incur high costs and thereby foreclose other, less expensive conservation options (Simberloff et al. 1992). For example, the Florida Natural Areas Inventory proposed a US$5 million corridor, in part to help protect the red-cockaded woodpecker (*Picoides borealis*). Some biologists predicted that this corridor would support only one successful woodpecker dispersal every five to seven years (Simberloff et al. 1992). The question, however, is not whether this seems to be a high figure. It is whether this is a cost-effective approach in comparison to other options such as translocation. Furthermore, cost-benefit analyses should take into account other corridor benefits such as maintenance or restoration of key ecological services. Finally, the cost of managing existing habitat, ecosystems, or landscape mosaics using the corridor approach is likely to be less than the cost of restoring these elements after their destruction. Options such as translocation fail to address any of these larger values that corridors can provide.

In summary, corridors are a potentially useful tool for maintaining and restoring biodiversity and ecological processes. However, empirical evidence for their purported benefits is either weak or lacking. Although recent studies have begun to demonstrate some of the benefits of corridors, much of that research fails to focus on issues of immediate relevance to conservation.

In the meantime, conservation practitioners have launched numerous corridor initiatives worldwide. This has put much of the largely theoretical debate on the merits and limitations of corridors to rest. What the conservation community needs now is further research about the effects of corridors in real landscapes. The ongoing initiatives are supplying useful insights about corridor design. Ultimately they will provide the real test of the hypothesized positive and/or negative effects of corridors.

3 Corridor Design

> The glorious mosaics of St. Mark's in Venice . . . appear as a
> pattern of colored patches and strips, usually with a background
> matrix. . . . The land appears much the same. . . . The individual
> trees, shrubs, rice plants, and small buildings, analogous to the
> tiny stones in the artist's mosaics, are aggregated to form the
> patterns of patches, corridors, and matrix on land.
> — Forman (1995:3–4)

There is no magic formula for designing biological corridors. Corridor design is highly specific to the unique requirements of the species, habitats, ecosystems, and ecological processes of concern (Friend 1991, Debinski and Holt 2000). Furthermore, design must account for not only the biophysical elements of a corridor but the socioeconomic and political factors influencing corridor configuration and implementation (e.g., Newmark 1993, Kaiser 2001). An understanding of corridor design can be obtained by reviewing how scientists and practitioners are approaching the problem, and by identifying guiding principles and recommending practical steps for the design process.

When considering approaches to corridor design, it is also important to bear in mind that corridors may not always be the most appropriate strategy for biodiversity conservation. While corridors are a useful tool for restoring connectivity and increasing effective reserve size, practitioners must assess their value based on the context and objectives of specific conservation efforts. The establishment of narrow, linear corridors should not be sought as a substitute for increasing the area of existing reserves: if used to justify the elimination of large habitat patches, for example, corridors could even undermine biodiversity conservation (Rosenburg et al. 1997). It is often better to preserve isolated sites of high conservation value than to acquire more accessible but marginal habitat for corridors (Noss 1987).

This chapter reviews key issues that can have critical implications for the design of corridors. It then examines how the objectives for a specific corridor provide a basis for design, and looks at elements key to corridor design.

The chapter concludes with an exploration of the practical steps involved in bringing corridor design from conception to implementation.

Corridor Context

Design strategies are highly dependent on the integrity of the landscapes in which corridors occur or are contemplated. In relatively disturbed landscapes, design should focus (at least initially) on linking habitat fragments, which may require building small-scale, linear corridors from scratch. On the other hand, in intact landscapes design involves protecting the predominant habitat matrix, which implies maintaining connectivity at the landscape scale. In the Y2Y region, for example, conservation planning in the southern Rocky Mountains focuses primarily on maintaining or restoring connections for wildlife across a developed landscape. By contrast, in the largely undeveloped northern Rockies, planning focuses on deciding where human activities can take place while still preserving maximum habitat connectivity.

Each of these approaches involves the two major corridor types described in chapter 2 (fig. 2.1). Landscape corridors tend to be appropriate in less disturbed landscapes where it is possible to maintain connectivity across large areas, such as in the Y2Y Corridor and the Klamath–Siskiyou ecoregion (cases 4 and 5, respectively). In contrast, on small scales and in relatively disturbed landscapes, linear corridors are more appropriate, such as the corridor designed for the golden lion tamarin (*Leontopithecus rosalia*) in a highly disturbed region of Brazil's Atlantic Forest (case 1).

Even at small scales in highly disturbed areas, however, a larger, landscape vision should guide the design of linear corridors. Ideally, they should be conceived of as building blocks for landscape corridors. In the absence of a clearly defined landscape vision, efforts to design and implement small-scale linear corridors — such as those along the Kinabatangan River (case 6) — are likely to be much less strategic. The Veluwe region of the Netherlands (case 8) provides an interesting case in which both corridor types are tightly integrated. Here a landscape corridor (which itself is nested within larger-scale corridors at the national and international levels) contains numerous linear corridors, all of which provide linkage through what are now highly disturbed or modified habitats and ecosystems.

As illustrated by many of the cases analyzed in this report, designing corridors only makes sense within a larger context such as ecoregion or landscape conservation. For example:

- Scientists have identified the Cascade–Siskiyou National Monument as a crucial link for wildlife moving between the Cascade and Siskiyou Mountain ranges (case 5).
- Conservation efforts planned for the 800 kilometer–long Terai Arc of the Eastern Himalayas are focusing on "priority bottlenecks." These efforts aim to conserve or restore forests and grasslands in critical areas, thereby increasing habitat connectivity for tigers (*Panthera tigris*) and greater one-horned rhinoceroses (*Rhinoceros unicornis*). Much of the existing habitat targeted for this purpose consists of multiple-use community forests (case 7).
- Along the Lower Kinabatangan River in eastern Sabah, Malaysia, restoration of riverine forests currently focuses on extremely small, linear corridors. To complement these efforts, WWF has begun to develop plans for a corridor that would extend from the Middle or Upper Kinabatangan to its mouth (case 6).
- Corridors in Canada's Bow Valley provide linkage for wildlife moving among several protected areas and other available habitat in the northern Rocky Mountains (case 4).
- A network of corridors in Florida, USA, is designed to connect Florida panther populations in protected areas located in the southern part of the state to habitat on private lands in central Florida and, ultimately, to protected areas farther north (case 3).

Ideally, a biodiversity vision for an ecoregion should be developed *before* beginning efforts to design corridors, so as to focus scarce resources on key biodiversity targets and in priority areas.

Corridor Objectives as a Basis for Design

Before one embarks on corridor design it is also imperative to consider what purposes or objectives the corridor needs to serve. Given scarce resources, it is often difficult to justify the establishment of corridors that serve biodiversity protection exclusively (Hellmund 1993). Whenever possible, corridors should be designed to achieve multiple objectives. In large-scale, landscape corridors, distinct objectives can be defined at different scales. For example, the Talamanca–Caribbean Corridor in Costa Rica is designed to protect habitats, watersheds, and movement for a wide range of species at a large scale, while also supporting ecotourism and agriculture at smaller scales (case 2). In the Terai Arc of the Eastern Himalayas, restored forest linkages provide harvestable resources and ecotourism income for local communities, while also preserving habitat connectivity and watershed values at a landscape scale (case 7). Wildlife corridors in Florida protect critical watershed values and support recreation and maintain habitat connectivity for Florida panthers and other wildlife (case 3).

If the purpose of a corridor is primarily to protect biodiversity, however, it is necessary to ensure that other uses do not undermine this objective (Noss 1991, Simberloff et al. 1992, Thorne 1993). In corridors designed for the Florida panther, for example, underpasses were constructed to permit crossings of road and highways (see box 5.2).

A corridor's objectives will play a critical role in guiding its design. This section examines how the objectives for a corridor—which range from enhancing animal movements over short distances to protecting entire communities and ecosystems—may guide the design process.

Protecting Focal Species

Worldwide the design of most corridors has been around focal species. Scientists and practitioners frequently use the distribution and habitat requirements of keystone or umbrella species to determine the size and configuration of biological corridors (see box 3.1 for definitions). This approach assumes that corridors meeting the needs of large carnivores or other far-ranging, fragment-sensitive species will effectively protect numerous other species. Use of a flagship species (box 3.1) can also help draw attention and build public support for corridor projects. Flagship species such as the golden lion tamarin in the Brazilian Atlantic Forest and the grizzly bear (*Ursus arctos horribilis*) in the U.S. Rocky Mountains are helping garner support for corridor initiatives (see cases 1 and 4).

BOX 3.1. TYPES OF FOCAL SPECIES USED IN CORRIDOR DESIGN

Keystone: A keystone species has a strong and unique impact on the ecosystem it inhabits. Removal of this species can disrupt ecosystem processes and possibly trigger the extinction of other species in the community. Large carnivores are often keystone species because their feeding behavior regulates the populations of other species.

Flagship: A flagship species is a charismatic animal that helps attract public support for conservation. Giant pandas (*Ailuropoda melanoleuca*) are a classic example of a flagship species.

---

continued

> *Umbrella:* Umbrella species require habitats and resources that also support a variety of other species, communities, and/or ecosystems. As a result, conservation efforts aimed at umbrella species are likely to generate broad conservation benefits. Umbrella species usually have (i) large area requirements; (ii) specific, well-defined habitat requirements; (iii) well-understood life histories (ideally subject to ongoing monitoring or study); and (iv) good chances of population stability or reintroduction to the area prioritized for conservation efforts. In the Y2Y Corridor, the grizzly bear is one of several umbrella species determining the design of landscape connectivity (case 4).

The design of a corridor intended to serve specific focal species requires knowledge of the species' basic ecology and life history traits, such as ecological role (e.g., predator versus prey), feeding behavior, habitat requirements, space-use patterns, and social organization. These traits help determine the likelihood that animals will find, select, and successfully pass through a corridor. For example, the probability of an animal *finding* a corridor depends on the distance it must travel to encounter the corridor and on its mobility and exploratory behavior. The probability that the animal will then *select* the corridor as a movement path depends on how it perceives the quality of the corridor habitat as compared to the patch and matrix habitats. Finally, the animal's ability to *traverse* the corridor depends on its finding required resources within the corridor while avoiding predators and other mortality factors (Rosenburg et al. 1997).

Migration patterns and the cues determining migration timing, direction, and distance can inform the design of landscape corridors. For example, the large-scale migrations of wildebeests (*Connochaetes taurinus*) and other herbivores motivated plans to link protected areas in Kenya and Tanzania. On a smaller scale, the migration patterns of elephants (*Loxodonta africana*) and other wildlife — and the economic importance of this wildlife — helped justify the partial removal of a fence between Botswana and Namibia (see box 4.1).

Many mammal species may be most likely to utilize corridors during dispersal, when young animals leave the maternal or natal home-range to establish their own territories (Harrison 1992, Beier 1995). Dispersal ecology — including dispersal timing, direction, and distance — can therefore provide useful insight for corridor design. Animals at high risk of predation, or that are very sensitive to human disturbance, are likely to require continuous habitat for dispersal and may utilize only corridors that are short relative to home-range size. Many group-living prey animals, such as white-tailed deer (*Odocoileus virginianus*), wild horses (*Equus caballus*), and dwarf mongoose (*Helogale parvula*) — which minimize predation risk by dispersing through short, rapid transfers between groups — appear to fit this model. By comparison, large carnivores tend to disperse much farther, traveling until they find a suitable territory. Few mammals, however, disperse more than five home-range diameters from the locale where they are born (Harrison 1992).

Knowledge of focal species' social organization can also shed light on dispersal patterns and corridor use, particularly in mammals. Depending on the social organization of a species, the presence of other members of that species may attract or repel individuals from corridors. Sex differences in dispersal patterns within a species also can have important design implications. For example, female tigers (*Panthera tigris*) and black bears (*Ursus americanus*) disperse only short distances and establish territories overlapping the maternal home-range. Corridors designed to accommodate female movements should therefore be wide and continuous enough to permit gradual range expansion across the landscape (Harrison 1992). Male tigers and black bears, by contrast, may disperse far beyond the natal range in search of new territory: longer corridors connecting distant patches might therefore be needed to accommodate male dispersal.

The feeding ecology of target species and other resource needs are also factors to consider when designing corridors. Lindenmayer and Nix (1993) emphasize that the presence of key food resources is essential when animals must spend extended periods in a corridor. Their studies of arboreal marsupials suggest that generalist feeders are likely to fare better in linear corridors than species that specialize on a dispersed or patchily distributed food resource.

Ultimately, appropriate corridor design may best be determined by combining knowledge of ecology and life history traits with empirical observation of actual space use by target species — as illustrated by a growing array of examples (box 3.2).

BOX 3.2. EXAMPLES OF APPLYING OBSERVATION
OF ANIMAL MOVEMENT, ECOLOGY, AND LIFE HISTORY
TRAITS TO CORRIDOR DESIGN

• To estimate theoretical minimum corridor widths for mammalian species, Harrison (1992) suggests using the home-range size of females.[a]

continued

- To determine a minimum corridor width for maintaining connectivity between upland and lowland habitats in the Usambara Mountains of Tanzania, managers used the minimum distance from the corridor's edge needed to find the most edge-sensitive bird species present (Newmark 1993).
- To estimate forest corridor widths for North American songbirds, practitioners have used the distance into the forest at which nest predation and parasitism become negligible (Newmark 1993).
- In southern California, researchers have tracked the nightly movements of radio-collared cougars to delineate range and travel routes (Beier 1992, 1993, 1995).[b]
- To identify priority corridor locations, scientists designing the Y2Y Corridor are using wide-ranging species that follow habitual movement routes — such as grizzly bears and bull trout (*Salvelinus confluentus*) (Herrero 1998; see case 4).
- Faced with insufficient data and few choices regarding corridor width and location, practitioners conserved a corridor linking upland and lowland habitats on Mount Kilimanjaro based solely on the observation that key focal species use it (Newmark 1993).

[a]Harrison (1992) has used this approach to estimate minimum corridor widths for wolves in Alaska and Minnesota, black bears in Minnesota, mountain lions in California, bobcats in South Carolina, white-tailed deer in Minnesota, and dwarf mongoose in Tanzania.

[b]Beier (1993) suggests that the cougar is a particularly useful focal species because its extensive nightly movements allow corridor routes to be mapped relatively quickly.

Protecting Communities and Ecosystems

While corridor design often revolves around focal species, it is important to understand that this approach does not guarantee protection of the full range of species and habitats of conservation interest. For example, in the Y2Y Corridor scientists found that corridors designed to serve grizzly bears in the Swan Valley of Montana would fail to protect key habitat for endangered bull trout (Wuethrich 2000). A more comprehensive approach that considers the needs of multiple focal species occupying a variety of habitats and niches is more appropriate when designing corridors to conserve biodiversity.

As discussed in chapter 2, corridors can serve a wide range of broader objectives. For example, they could protect and restore entire ecosystems and their services as a way to benefit both biodiversity and human needs (for example, fisheries, recreation, and agriculture), or ensure movement pathways for species and communities in response to global warming and other environmental changes. Box 3.3 discusses designing a marine reserve network to benefit both biodiversity and human needs.

BOX 3.3. MARINE RESERVE NETWORKS

Marine reserves have emerged as a promising strategy for both restoring fisheries and conserving biodiversity, and as a result they have expanded significantly since 1990. Such reserves can:

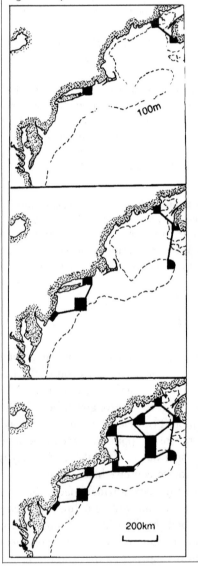

- help restock fishing grounds,
- protect vulnerable species,
- prevent or diminish habitat damage and promote habitat recovery, and
- maintain or restore overall biodiversity at the population, community, and ecosystem levels.

As is the case on land, no marine reserve is an island. Its effectiveness depends on the dispersal of marine organisms — which, for a large proportion of species, is one of the great mysteries of the sea. Almost all exploited marine species have a pelagic (or open sea) dispersal phase. During this phase, they release their eggs or larvae into open water, where they develop over periods of days to a few months. As a result, many of the species present in marine reserves are likely to have originated from somewhere else, which means that the reserves are unable to support self-sustaining populations of those species. The sources where reproduction of such species takes place are likewise unknown, although ocean currents probably determine the route over which dispersal occurs.

These issues raise special challenges for corridor design in marine environments. Since the source and the precise route of dispersal are in most cases unknown, patterns of ocean currents provide the best criterion for

continued

defining linkages within marine systems, and between those systems and adjacent freshwater and terrestrial habitats. This approach has been useful in designing large-scale linkages for coastal areas in and adjacent to the Mesoamerican Reef.

Another approach to conservation in marine systems involves establishing reserve *networks*, in which the distance between reserves is minimized — thereby increasing the chances for successful dispersal between them. This principle is illustrated graphically above: as the number of reserves (black rectangles) in a regional network increases, the connectivity between them (indicated by the number of links) increases even more rapidly.

— Courtesy of Roberts and Hawkins (2000)

To survive global climate change, plant and animal species will need to migrate across the landscape in response to shifting habitats. If climate change proceeds as quickly as current models predict, many species will not be able to migrate fast enough to survive. Corridors will at best act as a filter and may permit passage only by species with high mobility, broad dispersal, and generalist habitat needs. Conservation of large tracts of high-quality habitat in landscape corridors provides the best hedge against climate change impacts. Where large tracts of such habitat are not available, linear corridors, especially those along steep altitudinal gradients, are potentially the most effective way of enabling the biota to respond to climate change (Hobbs and Hopkins 1991).

Ideally, corridors designed to facilitate movement in response to climate change should form a network to allow migration in multiple directions (fig. 3.1). They should be as wide as possible to provide habitat as well as movement pathways for species with relatively low mobility, such as many plants and terrestrial invertebrates. Continuous habitat linkages along current climate gradients — both latitudinal and altitudinal — should be secured, although there is no guarantee that future climate gradients will parallel current ones. Prioritizing connectivity between areas that have served as refuges for biodiversity during past climate change may also be a good strategy (Hobbs and Hopkins 1991).

Corridor Design Elements

Most guidelines for corridor design offered in the literature focus on linear corridors and involve three interrelated structural elements: width,

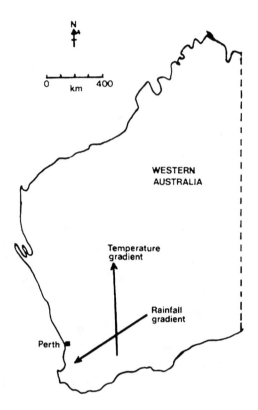

FIGURE 3.1

Factors in corridor design for climate change. Environmental factors are likely to change in different magnitudes and in different directions as climate changes. Since different environmental factors limit the distributions of different species, there may be a need for multiple corridors. These would allow all species in an ecosystem to shift their distributions to new suitable locations. Source: Hobbs and Hopkins 1991.

connectivity, and habitat quality (Thorne 1993). *Width* determines how much of the corridor interior is exposed to disturbances or edge effects, whether natural or human-induced, from the surrounding matrix. *Connectivity* refers to the degree to which gaps interrupt corridor habitat. *Quality* depends on both width and connectivity and reflects how closely the corridor approximates pristine habitat. Finally, other *landscape features* strongly influence each of these elements and their effectiveness in corridor design. Each of these elements is examined below, and a final section illustrates their application in practice and provides general guidelines for corridor design.

Width

Most research on corridor design has focused on width. The wider the corridor, the more interior habitat it will contain and the more protection it will provide for species sensitive to edge effects or disturbance from the surrounding matrix (box 3.4).

BOX 3.4. CORRIDORS AND EDGE EFFECTS

Linear habitats are especially vulnerable to disturbance, or edge effects, from the surrounding habitat matrix. This is also true for small habitat fragments as compared to large ones. A corridor that is not wide enough to provide an interior strip sheltered from edge effects may simply add more edge habitat to a fragmented landscape, leading to continued loss of native biodiversity (Harrison and Bruna 1999).

Studies highlight several specific edge effects associated with habitat fragmentation that can reduce corridor conservation value, including:

- increased microclimatic extremities;
- greater susceptibility to disturbances such as windthrows or fires;
- higher abundance of common, disturbance-tolerant species;
- release of predators or herbivores, which may destabilize food webs; and
- increased exploitation by humans (e.g., hunting, harvesting of plant resources, etc.).

In general, we can expect edge effects to be most severe when the corridor habitat and the matrix habitat differ greatly. Studies in tropical America document edge effects extending up to three hundred meters into forest habitat (Debinski and Holt 2000). Species that rely on interior habitat typically require corridors wide enough to provide a significant swath of habitat free from edge effects (Noss 1991). The impacts of edge effects on target species also depend on the sensitivity of those species to disturbance and on the amount of time they spend in the corridor (Newmark 1993).

The width required for a corridor varies according to the overall objectives and the needs of the target species. Scientists and practitioners generally agree that corridors designed for large, highly mobile species vulnerable to human disturbance should be as wide as possible. The most desirable width for a corridor may be moderated, however, by the need to discourage animals from lingering within its boundaries. For example, studies of root voles (*Microtus oeconomus*) suggest that corridors that are too wide may encourage the animals to pause for grazing, thereby increasing their potential exposure to predation (Soulé and Gilpin 1991). Especially in cases involving disturbance-sensitive animals in altered landscapes, practitioners may consider designing corridors of medium width. These would be neither so wide as to encourage wandering or pausing, nor so narrow as to discourage movement (Rosenburg et al. 1997). Another potentially moderating factor is the presence of large foraging animals or predators. Species such as elephants and tigers can cause conflict with people when they eat crops or prey on livestock. Corridors that are too attractive as habitat for these species may cause them to stay and become problem animals for nearby human populations.

Appropriate corridor width also depends on corridor length. The longer the corridor, the longer animals will spend traversing it and the more resources they are likely to need along the way. In fragmented habitats of southern California, for example, cougars (*Felis concolor*) disperse through corridors as narrow as 0.5 to 1.0 kilometers over a distance of 6 kilometers (Beier 1995). For long-distance travel, however, the corridor may need to be wider in order to accommodate foraging and resting in addition to movement (Harrison 1992, Beier 1995). This could be an important consideration for many species that are sensitive to human disturbance and require large foraging areas.

Corridor quality also influences the required width. Maintaining high-quality corridor habitat entails preserving a swath wide enough both to protect interior habitat from edge effects and to accommodate small-scale natural disturbance and succession (Noss 1991). As discussed below, actual observation of the movements of target species may be the best way to determine the habitat quality required.

In sum, corridor width, like most other aspects of corridor design, is highly case-specific.

Connectivity

The three major factors that determine a corridor's connectivity are: (i) the number and size of gaps in the corridor habitat, (ii) the presence of alternative pathways or networks, and (iii) the existence of larger habitat patches, or "nodes," along the corridor (fig. 3.2:D).

The degree of connectivity desired depends on the nature of the species and/or ecological processes that the corridor is designed to enhance. Many

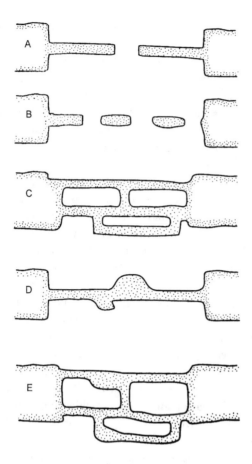

FIGURE 3.2

Factors affecting corridor connectivity. A corridor's connectivity is influenced by various factors, including (A, B) the length and number of gaps, (C) the presence of a network of multiple
linkages, and (D) the presence of nodes or patches of habitat. A corridor that provides continuity, multiple linkages, and associated nodes of habitat is likely to maintain maximum
connectivity (E). Courtesy of Bennett (1999).

species that are sensitive to human disturbance, such as cougars, appear to
require continuous habitat cover for movement (Beier 1993). Gaps of low-
quality habitat may also lead to high mortality for less mobile species or
those vulnerable to changes in microclimate, such as salamanders (Rosenburg et al. 1997). Other species normally intolerant of humans can cross
some types of gaps relatively well: for example, the highly endangered golden

lion tamarins in the Atlantic Forest are capable of dispersing across highways (see case 1).

In some cases, continuous corridors may not be necessary or even desirable. Some species, including endangered British red squirrels (*Sciurus vulgaris*), can use stepping stones, or multiple fragments, of habitat located within a critical distance to disperse and breed across considerable areas (Hale et al. 2001; fig. 3.2:B). Some disturbance-sensitive species, such as the northern spotted owl (*Strix occidentalis*), may actually move more safely through stepping stones of forest habitat than through narrow, continuous corridors. Spotted owls disperse in random directions, reducing the likelihood that they will find and utilize linear corridors. They may also experience increased predation in narrow, linear habitats (Thomas et al. 1990, Forsman et al. 2002).

The presence of alternative pathways or networks between suitable habitats appears to be another important factor determining the degree of connectivity in a corridor (fig. 3.2:C). Multiple pathways build a degree of redundancy, thereby reducing the risk of losing overall connectivity due to habitat conversion or alteration at a specific locale. Multiple connections can also increase the likelihood that target species will find and use the corridors (Noss 1991, Beier 1992). This would appear to be the case for the northern spotted owl, as mentioned above (Thomas et al. 1990).

A final factor influencing corridor connectivity is the existence of habitat "nodes" along its length (Noss and Harris 1986; fig. 3.2:D). Nodes may consist of protected areas or simply larger habitat fragments that have not been converted. They provide resources to support individual animals, populations, communities, or ecological processes along the corridor length. Nodes become important when animals spend so much time in the corridor that they need to forage or take shelter. The desired area of nodes, and the resources they contain, depends on which species they are expected to support and for how long.

Habitat Quality

Closely related to corridor connectivity is the issue of quality, which refers to how closely corridor habitat approximates intact or interior habitat. To sustain multiple species or entire biological communities across a landscape, corridor habitat generally should include intact ecosystem structure, maximum levels of native species diversity, and minimal intrusion by edge or exotic species (Noss 1991). Corridors designed to provide habitat or facilitate movement of wildlife should provide the highest-quality habitat possible for the most sensitive species targeted for conservation. As with corridor connectivity, the quality required may vary greatly among species and is best determined through empirical observation of animal movement patterns. Key elements of habitat quality include vegetative cover, topography, and level of human disturbance.

In relation to vegetative cover, it is important to consider what purpose the vegetation serves. If animal movement is the primary goal, does vegetation need to provide food resources or protection against microclimatic variations or predators? Maintenance or restoration of native vegetative cover is a prime requirement for habitat effectiveness. Some species may choose increased vegetative cover for traveling, even at the expense of other resources. For example, when crossing highly modified landscapes the endangered Iberian lynx (*Lynx pardinus*), which is highly mobile and intolerant of humans, uses habitat with lower prey density but higher understory cover than its preferred habitat (Palomares 2001).

Landscape Features

Landscape features strongly influence the design of corridors, and they provide clues to appropriate corridor location. Studies of arboreal marsupials in Australia suggest that corridors connecting diverse landscape features may harbor more biodiversity than do corridors linking just one type of feature (Lindenmayer and Nix 1993). In addition, some types of features may channel or facilitate animal movement. Cougars moving through fragmented habitats of southern California, for example, favor covered ridges and the scour zones of streambeds for travel, and they also utilize lightly used dirt roads and hiking trails to traverse dense brush (Beier 1995).

River and stream systems are frequently a fundamental component of biological corridors because they tend to concentrate biodiversity and serve as avenues for the movement of organisms and materials (Noss 1991). In fact, they can serve all the essential functions characteristic of corridors — including habitat, conduit, barrier or filter, and source or sink (fig. 2.2). As part of a landscape-level conservation plan, riparian corridors can also provide a cost-effective way to preserve watersheds and aquatic ecosystems that benefit human communities (Bennett 1999). In short, riparian corridors provide strategic building blocks for designing and implementing both linear and landscape corridors.

To conserve biodiversity and environmental services, riparian corridors should maintain or restore a swath of natural vegetation wide enough to incorporate the geomorphic floodplain, and the headwater and groundwater sources that feed the waterway. Sites in special need of healthy riparian vegetation include steep slopes, deforested or overgrazed areas, and croplands — all of which can be a major source of sediment and runoff. Aquifer recharge and discharge zones are also important because they keep streams flowing year-round. Ideally, a riparian corridor should extend beyond the geomorphic floodplain to provide a long-term sink for excess sediment and nutrients (Binford and Buchenau 1993).

BOX 3.5. ROADWAYS AND CORRIDORS

Because they provide networks of continuous passageways traversing large areas, existing roadways appear to offer important opportunities for enhancing landscape connectivity. Roadsides can serve as habitat for wildlife, and they have allowed some small mammals and edge-tolerant bird species to expand their geographical ranges. Several species of large mammals — including wolf (*Canis lupus*), dingo (*C. familiaris dingo*), cheetah (*Acinonyx jubatus*), lion (*Panthera leo*), and African elephant (*Loxodonta africana*) — utilize lightly traveled roads for passage. Roads are most likely to benefit wildlife movement when they receive light use and include a broad swath of intact habitat on either side (Bennett 1991).

The potential for roadways to serve as biological corridors is limited, however, by hazards that also make roads formidable barriers to wildlife movement. Roads attract some animals, which can then experience high roadkill mortality. Roadkill is the leading cause of death for several large mammal species in Florida, including Florida panthers, black bears, and Key deer (*Odocoileus virginianus clavium*) (Harris and Gallagher 1989). Roads through intact habitats enhance human access and can lead to increased deforestation, as is the case in the Brazilian Amazon. Roadways may also serve as conduits for exotic or invasive species (Bennett 1991). For these reasons, corridors for most species should avoid crossing roads whenever possible (Noss 1991).

Where roads are unavoidable, under- and overpasses may help facilitate passage by some animals. Underpasses have proven beneficial for mountain goats (*Oreamnos americanus*) in Banff National Park (case 4), for red deer and wild boar in the Veluwe region of the Netherlands (case 8), and for panthers in Florida (case 3) and southern California (Beier 1995). Other ways to reduce conflicts between roadways and wildlife include installing wider bridges over ravines and waterways, installing fencing to keep animals away or signs to warn drivers, and closing roads seasonally (Bennett 1991).

Design Application and Guidelines

Scientists have identified some general design guidelines that can increase the biological value of corridors and help minimize the potential negative impacts discussed in chapter 2. We give a brief summary here, but see Harris and Scheck 1991, Noss 1991, and Thorne 1993 for more detailed discussion.

- Link only patches that were formerly connected and that contain naturally contiguous habitat types. This should help to avoid unnatural range expansion or introduction of invasive species to patches of high-quality habitat.
- Minimize connection of artificial or disturbed patches to higher-quality habitat.
- Identify and preserve existing natural corridors such as riparian zones and migration routes. Riparian zones can often help protect water quality and maintain high concentrations of biodiversity, particularly in arid regions.
- Place corridors along altitudinal and latitudinal gradients to incorporate maximum biodiversity and mitigate the effects of climate change.
- Avoid long stretches (>2 km) without nodes, and build redundant connections via alternative pathways or networks.

Since the early 1990s researchers have developed more specific guidelines for corridor design. One of the most sophisticated and detailed sets of guidelines was created for the Bow Valley in the Y2Y Corridor (case 4). These provide a basis for mitigating development impacts in one of the few remaining linkages for birds and mammals in the southwestern Canadian Rockies (see box 5.6). Based on research and conservation biology theory, these guidelines provide formulae for relating variables of length, width, shape, topography, vegetative cover, and adjacent land uses in corridor and patch design. Another example of the application of the various design elements described above is a nationwide network of forest corridors in Bhutan (box 3.6).

BOX 3.6. DESIGNING A NATIONAL CORRIDOR NETWORK IN BHUTAN

In Bhutan, a high degree of intact habitat remains relative to other parts of the Eastern Himalayan region. While approximately 25 percent of the country's territory currently is under protection, the government has committed to preserving a minimum of 60 percent under forest cover in perpetuity. The establishment and maintenance of corridors linking existing protected areas would play a key role in achieving this ambitious goal.

---

continued

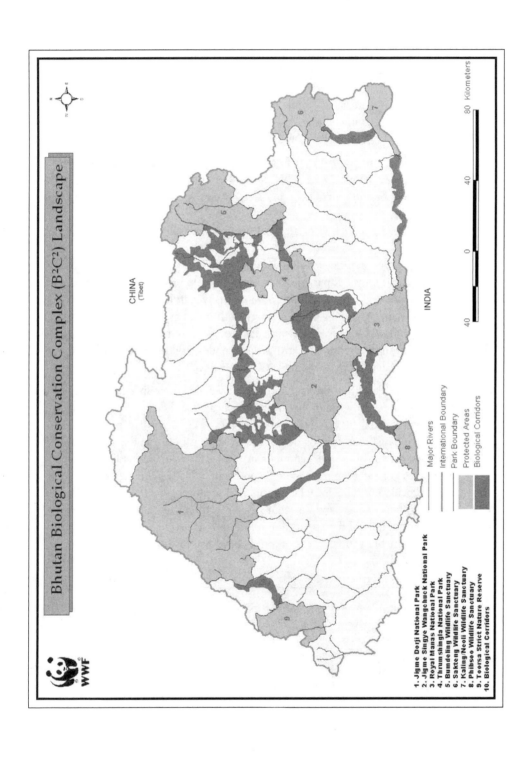

Bhutan Biological Conservation Complex (B²C²) Landscape

CHINA
(Tibet)

INDIA

Major Rivers
International Boundary
Park Boundary
Protected Areas
Biological Corridors

1. Jigme Dorji National Park
2. Jigme Singye Wangchuck National Park
3. Royal Manas National Park
4. Thrumshingla National Park
5. Bumdeling Wildlife Sanctuary
6. Sakteng Wildlife Sanctuary
7. Kaling/Neoli Wildlife Sanctuary
8. Phibsoo Wildlife Sanctuary
9. Toorsa Strict Nature Reserve
10. Biological Corridors

40   0   40   80 Kilometers

WWF

WWF developed a preliminary plan for a national system of corridors in Bhutan. The process of selecting land for corridor designation involved a scoring mechanism in which land was rated for suitability according to specific criteria, including:

- abundance of target wildlife or their prey (focal species),
- width of the narrowest constriction (corridor width),
- movement of key wildlife species (connectivity),
- condition of habitat canopy and undergrowth (habitat quality),
- levels of human disturbance (habitat quality),
- occurrence of forest fires (habitat quality), and
- slope of terrain (landscape features).

As a result of this scoring matrix, the areas selected as paths for the Bhutan corridor network are at least 1–3 kilometers wide and are characterized by an intact forest canopy, minimal human disturbance, and terrain that is flat to moderately steep. Selection criteria also included evidence of the movement of key wildlife species. The plan revealed several priorities that will be critical for achieving the overall conservation goal. For example, ThrumshinghLa National Park (no. 4 on map) lies at the heart of a contiguous habitat for the Bengal tiger (*Panthera tigris tigris*) and is critical for conservation of that species. Because of its location in the central part of Bhutan, the park will serve as a "hub" for Bhutan's entire corridor network.

—Adapted from WWF Bhutan Program: (http://www.wwfbhutan.org.bt/)

Steps in Corridor Design

Information Needs

Taking into consideration the aspects of corridor design described above, we can define the major steps involved in bringing corridor design from conception to implementation. Before we go through these steps, however, it is important to consider the range of information needed for designing a corridor — which is potentially immense and can lead to paralysis through analysis. To avoid this situation, it is critical to be strategic about information needs at different scales, and to focus on different scales at distinct stages of the design process (box 3.7).

Although increasingly detailed information becomes desirable as design progresses, often resource managers must make decisions based on very limited data. Furthermore, very specific information is often the most expensive and time-consuming to acquire. This type of data should be sought only once its usefulness in designing and implementing the corridor is known. It is not always necessary or advisable to wait for completion of in-depth surveys for very specific information — such as the identity of landowners and their property boundaries — before beginning to select candidate areas and build stakeholder participation. Instead, data gathering used for corridor design should proceed from coarser scales, where general and qualitative data are appropriate, to finer scales, where specific and quantitative data work best.

Finally, because corridors and other landscape elements are dynamic, it will be necessary to define and monitor a few strategic indicators relevant to corridor integrity — especially those corresponding to connectivity and habitat quality — as a basis for making future adjustments to corridor design.

BOX 3.7. DESIRED CHARACTERISTICS OF INFORMATION
GATHERING FOR CORRIDOR DESIGN

Ideally, information for corridor design should be:

Systemic, in order to deal with the landscape as an interrelated whole rather than as isolated parts;

Scale sensitive, moving from general data at the outset to more detailed information as corridor sites become defined;

Adaptable, so that the design process can be adjusted to meet unique local conditions;

Concrete, addressing tangible and specific objectives rather than abstract values and goals;

Frugal and flexible in the use of data, to make good use of limited data and permit revision as new information becomes available;

Useful, as a basis for defining a few strategic indicators to be used for monitoring and making necessary adjustments to corridor design over time; and

Participatory, to address the needs of diverse stakeholders.

— Adapted from Hellmund 1993

Below we outline four steps in corridor design (adapted from Hellmund 1993). These steps provide insight into the kinds of challenges — both technical and "nontechnical" — that are likely to emerge in designing corridors. The steps are not written in stone, and in some cases (especially those involving smaller-scale, linear corridors) a "bottom-up" approach — proceeding from small to larger scales — may be preferable. Nevertheless, an approach that proceeds from larger to increasingly finer scales has important advantages, especially in defining how localized initiatives relate to broader conservation objectives at the ecoregion level, and in determining appropriate information needs.

Step 1: Understand the Broad Conservation Goals and the Regional Context

As discussed previously, the process of designing corridors begins with the conservation targets defined as part of a large-scale conservation or ecoregional vision. Because corridors provide the operational means of achieving those targets, the first priority at this stage is to define the approximate scale and type of corridor needed. The second priority is to identify special opportunities and threats that will help in designing and implementing the corridor — or that could potentially undermine it. For example, the conservation vision for the Eastern Himalayas ecoregion aims to protect, restore, and connect formerly linked ecosystems. The Terai Arc offers special opportunities for linkage because of the existence of a string of protected areas along its 800 kilometer–long axis, and because of the emergence of incentives for communities to adopt sound forest management practices outside those areas (see case 7).

In addition to an ecoregional vision, at this stage the best sources of information are likely to be residents or technical experts familiar with the region and with the key trends involving resource use and conservation (e.g., land-use change, economic activities, population growth, etc). Large-scale maps denoting the "lay of the land," including major geographic features and population centers, are probably sufficient at this stage.

Step 2: Identify Specific Corridor Objectives and Define General Corridor Areas

The next step is to define what objectives the corridor should address. The objectives should be as specific as possible. For example, instead of a vague objective such as "enhancing wildlife movement," the corridor could provide habitat or movement corridors for a specific species or group of species. What other objectives could a corridor achieve? How do the objectives interact and possibly conflict? For example, if recreation is an objective, how can it be reconciled with biodiversity preservation?

After defining objectives for the corridor, it is important to identify areas that are potentially most appropriate for corridor development. Ideally there will exist more than one candidate area, thus permitting a choice based on which area offers the best conditions. Deciding which areas are appropriate will involve analyzing existing or potential linkages among habitats, ecosystems, and ecological processes. Candidate areas could be identified according to a variety of criteria, as discussed above. Appropriate candidates might, for example, be those in which corridors could link habitats over minimum distances and/or areas with sightings of target species. Alternatively, the most appropriate areas might be those that could accommodate numerous corridor objectives.

To determine specific corridor objectives and potential locations for achieving them requires specific information about the biological, hydrological, ecological, and/or recreational assets that a corridor could enhance. Appropriate information sources are likely to include people knowledgeable about specific areas within the region, as well as maps showing the distribution and condition of protected areas and natural resources.

Step 3: Select a Candidate Area and Determine a Tentative Corridor Configuration

At this stage much more detailed information is required on the candidate areas under consideration. A good strategy here is to look for priority areas where existing corridors already provide connectivity, such as gallery forests or habitat fragments that serve as "stepping stones" for movement of biota. Data likely to be useful will reveal movement patterns and other life history traits of specific target species as they relate to the ecological features (e.g., resources, disturbance levels, topography, etc.) of candidate areas. If such specific data cannot be obtained, more general information can be helpful — such as data on the use of riparian corridors as conduits for biodiversity, or the condition of habitat linkages along gradients in soil, habitat structure, altitude, and climate.

Social, economic, and political factors are also likely to become critical at this stage and will largely determine the long-term success of the corridor. It is important to consider how factors such as existing land uses, tenure arrangements, administrative boundaries, local governance, and policies will affect corridor implementation. Furthermore, it is essential to understand how a corridor will impact different stakeholder groups, the attitudes of stakeholders toward corridor development, and what type and level of incentives are likely to promote cooperation.

At this stage consultation with a wider range of stakeholders usually becomes a central part of corridor design, and as a result the design process becomes less of a technical exercise and more of a social and political one (WWF 2000). If consultation is an integral part of decision making, the outcome of the design process will be more open-ended and less predictable. Conflict resolution frequently becomes a critical tool for diverse interest groups to negotiate mutually acceptable solutions (WWF 2000). Without sufficient stakeholder buy-in, corridor implementation is much more likely to fail.

Step 4: Define Strategic Interventions

As mentioned above, socioeconomic and political issues, as well as the scale and ecological conditions characteristic of the region, will determine appropriate approaches to corridor implementation. For example, in the case of a small, linear corridor, a straightforward and relatively interventionist approach of acquiring property or property development rights (easements), or of enforcing zoning codes, may be appropriate. Implementing a landscape corridor, in contrast, is a regional planning exercise that will probably require more carrots (i.e., incentives) than sticks (i.e., enforcement). As discussed in chapter 4, a participatory approach is essential to establish the legitimacy of a corridor design.

One practical way to approach implementation is to design linear corridors as integral components of a larger landscape corridor. A plan for a landscape corridor identifies regional goals that are achievable over the long term and provides a context for devising and implementing appropriate policies, while linear corridors serve as locales in which local stakeholders can achieve tangible successes over a shorter term. This approach allows practitioners to strike a balance between more general, long-term objectives (such as those associated with a landscape corridor) and more localized, short-term objectives that can motivate stakeholders (such as those associated with linear corridors).

A number of case studies presented in chapter 5 illustrate the use of this approach. In Florida, for example, The Nature Conservancy (TNC) prioritizes land acquisitions based on the goal of creating a network of continuous habitat extending across the state. With this long-term goal in mind, TNC focuses on designing and establishing smaller-scale, linear corridors as opportunities arise (case 3). Similarly, the Terai Arc provides a landscape-scale framework for preserving habitat connectivity across the Eastern Himalayas. By examining conservation needs and opportunities across the whole arc, practitioners have identified priority bottlenecks where implementing corridors in the near-term will provide the greatest benefit to local communities and overall landscape connectivity (case 7).

4 Corridor Implementation

This chapter examines five key questions related to corridor implementation. The answers to these questions are likely to determine whether corridor implementation can begin in the first place, and whether it will be sustainable over the following years and decades. Following this discussion, the chapter concludes with an analysis of how issues such as incentives, resource management, and governance need to be addressed together as part of a corridor implementation strategy.

The questions are:

How do you manage corridors?
What are the obstacles to implementing corridors?
How do you build support for corridors?
What are effective incentives for corridors?
How should corridors be governed?

How Do You Manage Corridors?

The primary conservation objectives largely determine the management options for biological corridors. We consider three major management objectives below with illustrations from the case studies (chapter 5): facilitating animal movement, maintaining or restoring environmental services, and maintaining or restoring habitats and ecosystems.

Facilitating Animal Movement

The case studies later in this book provide diverse examples of how corridors are managed to facilitate animal movement. Achieving this objective requires a solid understanding of the focal species, and specifically of the kinds of resources and habitat it requires and when. Corridors for specialist species requiring extensive areas of pristine habitat (such as grizzly bears) present extremely challenging management demands, whereas corridors for more generalist species (such as elephants) or species requiring smaller range areas (such as anteaters) are easier to manage.

The most straightforward approach to facilitating animal movement involves removal of artificial impediments across the predominant line of movement. Dams are being disassembled in the Little Applegate Watershed and elsewhere in the Klamath–Siskiyou ecoregion as a way of facilitating salmon migrations. The decision to remove dams typically involves long and complex negotiations, but it reflects a growing public perception of watersheds as a source of environmental services other than power generation or spaces for flood control.

A more sophisticated approach to facilitating animal movement is to build corridors over or under human-constructed impediments such as highways. Illustrations of this approach are the ecoducts for deer and boar movement in the Veluwe (see fig. 4.1), and the underpasses for panthers in Panther Glades. Ecoducts and underpasses are expensive, and a key question involves their cost-effectiveness — an issue discussed below in the Veluwe case.

The most complex approach — which is strongly associated with the origins of the corridor concept — is to manage *habitat* for animal movement. At one extreme, such management involves the use of short, linear corridors, such as those used to connect the highly fragmented habitats of the black and golden lion tamarins. Key questions here are: How wide a corridor is needed to allow tamarins to move and to protect them from edge effects such as wildfires, and how wide can gaps in the corridor be without impeding movement? In relatively short corridors such as these, one major habitat type is usually involved — thereby greatly simplifying management.

At the other extreme, managing for animal movement can involve wide-ranging species — such as elephants (Kinabatangan), tigers (Terai Arc), and grizzly bears (Y2Y) — over immense landscapes. This situation is much more complex because it involves a spatial and temporal patchwork, or mosaic, of

FIGURE 4.1

Ecoduct in the Veluwe region of the Netherlands. Courtesy of Vreugdenhil 2000.

habitats and ecosystems, some of which may be in a "natural" condition and others of which may be highly altered. It also involves a much wider range of social actors and their interests, incentives for conservation, and governance arrangements. For these landscape corridors, knowledge of several focal species is often required. Corridors designed to maintain connectivity for entire biological communities — such as those designed to minimize the negative impacts of climate change on habitats and ecosystems — are the most complex of all.

Maintaining or Restoring Environmental Services

Integrated concepts of ecosystem management are beginning to broaden the vision of resource managers beyond market commodities to an array of environmental goods and services that ecosystems provide for free. Many of those services are associated with water, and new financial instruments for sound management of watersheds provide promising incentives for conservation (Johnson et al. 2001). Examples of water-related services that corridors could potentially provide follow.

- Along the Kinabatangan River, some oil palm companies have established pilot plantations of economic tree species as an alternative way to generate income while reducing damage by floods and elephants (*Elephas maximus*).
- In Panther Glades, restoration of natural hydrological regimes will provide quality sources of water for urban centers in southern Florida.
- In the Cascade–Siskiyou National Monument, road closures and restrictions on logging and grazing are being implemented to maintain watershed quality and habitat, primarily for fishing.

A contentious issue is whether managing for environmental services requires the maintenance or restoration of the full biotic richness of an ecosystem. For example, native riparian vegetation may not provide the only viable means of protecting water resources. For some greenway projects managers may favor exotic vegetation that is hardy, attractive, and easy to maintain. Nevertheless, natural vegetation is often the most effective for protecting water quality, and once established often requires little or no maintenance (Binford and Buchenau 1993).

Management for environmental services can and should be compatible with biodiversity conservation as an important secondary objective.

Maintaining or Restoring Critical Habitats and Ecosystems

As discussed in chapter 2, maintaining intact habitats or ecosystems is more cost-effective than their restoration. Likewise, except under rare situations, natural restoration is more cost-effective than artificial restoration. Biological corridors are often established as a way to preclude or deflect destruction of habitats, ecosystems, or ecological processes. This is particularly true in places where relatively intact habitats or ecosystems still remain — such as in the Klamath–Siskiyou ecoregion, the Y2Y Corridor, and the Talamanca corridor. Here there are still opportunities to set aside pristine spaces that can also contribute to connectivity.

In more altered sites or landscapes, the next-best option is to restore habitats or ecosystems — preferably by natural means. Examples of this approach follow:

- In the Terai Arc, encouraging natural regeneration and limiting extraction has helped restore degraded forests in priority bottlenecks.
- In Panther Glades, prescribed fire is a low-cost way to restore heavily logged pine forests.

Artificial restoration can work where other options fail or where other considerations besides costs prevail. For example:

- In the Cascade–Siskiyou National Monument, the closure of roads and the planting of riparian vegetation are two ways to restore watershed and stream habitat.
- In the Veluwe, restoration of some forestlands requires the establishment of plantations, and removal of contaminated topsoil — while costly — effectively restores heathlands.

What Are the Obstacles to Implementing Corridors?

We examine five major groups of obstacles below:

- threats to biodiversity,
- lack of awareness, understanding, or concern,
- resistance to decreased resource control and use,
- corridor costs, and
- their uncertainty and complexity.

Threats to Biodiversity

As discussed in chapter 1, the most visible sign of ecological destruction is the fragmentation of natural habitats and landscapes — the major cause of species extinction today. Biological corridors are designed to deflect, diminish, or reverse such fragmentation, but they do not address its root causes. These causes frequently reflect fundamental socioeconomic processes — such as population growth, infrastructural development, and conflict — that are extremely difficult to reverse (Wood et al. 2000). Without other efforts to address these causes, corridors alone are unlikely to be an effective conservation strategy over the long term. The primary purpose of Panther Glades corridor, for example, is to expand panther habitat, but it does not in itself decrease development pressures in southern Florida. The Cascade–Siskiyou National Monument secures a corridor for wildlife movement between the Cascade and Siskiyou mountain ranges, but logging and urban sprawl in adjacent lands may eventually make this corridor a habitat island.

The need to address root causes of biodiversity loss provides yet another reason for designing and implementing corridors as part of a larger-scale biodiversity vision.

Lack of Awareness, Understanding, or Concern

The cases presented in this book emphasize the critical need for engaging a wide range of stakeholders to build support for biological corridors. The lay public has little awareness of the need for biological corridors and landscape connectivity. Lack of understanding is reflected by the considerable scientific uncertainty surrounding corridors, and by the dearth of information concerning their socioeconomic dimensions. Lack of concern is reflected by the fact that biodiversity conservation often takes a back seat to other social issues, such as development and poverty alleviation. Indeed, the major problems that corridors are designed to address — habitat fragmentation and resulting biodiversity loss — are issues often difficult to perceive, especially at the large geographic scales where broad public support for corridors is most critical.

Resistance to Decreased Resource Control and Use

Corridors usually extend outside of parks and reserves, traversing areas owned and/or used by private landowners or communities. These groups tend to be suspicious of corridor initiatives, which involve multiple interest groups — many of whom may be outsiders — and often require some form of governmental intervention. To the degree that they restrict resource control and use, corridors generate proportionately greater suspicion or even active resistance. For example, many ranchers in the Y2Y region are highly suspicious of conservation efforts that may impinge on their grazing practices or force them to accommodate large predators that threaten livestock. Likewise, landowners in the Talamanca region were initially reluctant to accept new land-use restrictions, especially any that might limit the harvesting and sale of timber. In Florida, landowners are frequently unwilling to sell their land or development rights (i.e., easements). Property owners tend to be extremely suspicious of stricter zoning that may be associated with corridor initiatives, as is apparent in the case of the Veluwe region in the Netherlands.

Resistance to corridors that impinge on local rights to control and use resources appears to be greatest in places where those rights are most clearly defined. Resource rights are not limited to private property and, in the case of corridors, often include public or common property. In the Muskwa–Kechika area of the Y2Y Corridor, for example, petroleum, timber, and mining companies reacted strongly against restrictions to publicly owned lands. Although the first two groups reached a compromise that would secure their future operations while maintaining landscape connectivity, the mining industry pulled out of the negotiation process. Likewise, in the Klamath–Siskiyou ecoregion, increased restrictions on use rights by ranchers and loggers on public lands have generated strong opposition from these groups.

In the Terai Arc, designating new protected areas on public lands would deprive local communities of such subsistence resources as nontimber forest products. For this reason, instead of establishing new protected areas, groups involved in corridor implementation are seeking to strengthen environmentally sound livelihoods such as community-based forest management as an incentive for conserving forest linkages. Ironically, however, one of the first protected areas in this corridor — the Chitwan National Park — was created in part by relocating local villages that were in constant conflict with a resident tiger population. This measure helped reduce human-wildlife conflicts and provided an opportunity for people to obtain better land. Yet only under highly specific conditions is relocation likely to be advantageous for both people and conservation.

Resistance to decreased resource control and use is not limited to the private sector. For example, a county government in Florida's Panther Glades corridor initially objected to transferring private land to state control for fear of losing revenue from land taxes. The issue of public control over corridors, which becomes even more complex at an international level, is examined in the section "How should corridors be governed?" below.

Corridor Costs

Because few biological corridors have been fully implemented, there is almost no information about the costs of their establishment and maintenance. Of the cases examined in this book, the only precise data involve the costs of overpass crossings of major highways to permit passage of wildlife in the Netherlands (approximately US$4–5 million each: see fig. 4.1), of underpasses of highways to permit passage of wildlife in southwest Florida (ranging from US$128,000 to $350,000 each: see box 5.2), and of land purchases for priority panther habitat in southwest Florida (which average US$4,700–7,400 per hectare for privately held areas with agricultural potential; see Main et al. 1999).

Some cases revealed that the costs of artificial restoration are prohibitive except under special conditions. In the Veluwe, for example, restoring a degraded heathland can cost over US$5,000 per hectare.

Land for corridors may be more expensive than land acquired for protected areas in the past. This is because corridors often connect existing protected areas — many of which exist in part because they had low development value — by incorporating land that is attractive for other economic uses. Nevertheless, the important linkage role that corridors provide can increase their appeal as conservation investments and open up new possibilities for funding.

Lowering the price of land acquisitions can reduce substantially the costs of establishing corridors. For example, instead of outright purchases of property, it might make more sense to purchase development rights or easements — which usually cost far less. This strategy has been used successfully by The Nature Conservancy in the United States and, increasingly, in Latin America.

Like environmental initiatives in general, corridor projects face formidable funding needs. In the Veluwe, corridor initiatives are estimated to require US$9 million (8.1 million euros in mid 2004) during 2001–2010.[1] According to the plan, funding is to be provided by public sources at local, national, and international levels. Fundraising needs have been a major impediment to corridor establishment in the Pinhook and Panther Glades areas of Florida. In addition, corridors represent a long-term investment that will require prolonged if not permanent commitment by key stakeholders — which in today's world of rapidly shifting priorities is extraordinarily rare. The case studies in this book show that funding for corridors comes from diverse sources, such as donations and grants in Terai Arc and Klamath–

Siskiyou, state land programs and water management districts in Florida, and local municipalities, provincial governments, and trust funds managed by stakeholders in Y2Y.[2]

While there is little information on corridor costs, there is even less on the costs of *not* maintaining or establishing corridors. One example we do have involves the West Caprivi Strip between Namibia and Botswana, where the fencing off of a critical migration route was estimated to generate substantial losses of revenue in tourism and related activities (box 4.1). These losses appear to be many times greater than the problem that the fencing was designed to address. Another study by Binford and Buchenau (1993) showed that restoration of riparian vegetation can be much less expensive than other options, such as building new filtration facilities.

BOX 4.1. IMPLICATIONS OF CLOSING A MAJOR WILDLIFE
MIGRATION CORRIDOR: THE CASE OF THE WEST CAPRIVI STRIP
IN SOUTHERN AFRICA

The West Caprivi Strip lies on the border between Botswana and Namibia and is home to Namibia's richest diversity of wildlife (see fig. 4.2 below). Most of its mammalian and avian species are migratory. During the dry season, wildlife congregates on the Kwando and Okavango Rivers, grazing in productive floodplains or browsing in the adjoining Kalahari woodlands. In the wet season, wildlife moves toward Namibia. Because of these processes, the West Caprivi Strip is the most environmentally sensitive area along the entire Botswana-Namibia border.

Following a severe outbreak of bovine lung disease in both Namibia and Botswana in 1996, the government of Botswana eradicated 200,000–300,000 head of cattle, compensated affected ranchers, and constructed electric fencing along the country's border with Namibia. Beginning at the Okavango River in the west, the fencing extended along the entire West Caprivi border area (190 kilometers) to the Kwando River in the east. The barrier consists of two identical barbed-wire fences. Vegetation between and adjacent to the fences was bulldozed in a 200 meter–wide strip.

Because of the West Caprivi's importance as a migratory route, the fencing poses a significant threat to wildlife. While some animals manage to get through, most turn back (Weaver 1997). Among the most affected animals are elephants (*Loxodonta africana*), whose numbers — 8,831 were counted in 1994–1995 (Conservation International 2001) — probably would undergo a substantial decline if the fencing were maintained as originally constructed. Especially

continued

during the dry season, wildlife becomes separated from watering holes and animal populations face severe stresses. Over the short term, the fences can ensnare numerous animals. Over the mid-term, destruction of riverine habitats could result from overgrazing by elephants and other animals. Eventually, population crashes of many wildlife species in the West Caprivi could take place (Weaver 1997).

The economic costs of the disease and Botswana's responses to it were considerable. The country's domestic beef production dropped from 25.5 million tons in 1996 to 12.7 in 1998 (FAO statistical database). The country's trade balance changed as beef exports declined by US$2.4 million and imports increased by $2.3 million. Yet the response appears to have been disproportional to these costs: the government spent a total of US$42 million in response measures, including $32 million on detection, eradication, and compensation; $7 million on contracting of experts and consultants; and $3 million on fence construction. In addition, the value of wildlife must be factored into a cost-benefit analysis. A 50 percent loss in wildlife abundance could generate losses of US$3.3 million in Botswana's total revenues from tourism (Weaver 1997).

Given these considerations, it is striking that the fencing was placed in an area where commingling of Botswana and Namibia cattle is minimal and the impacts on wildlife are greatest. Following lengthy discussions with Conservation International (CI) and WWF, the Botswana government agreed to remove the easternmost 30 kilometers of the 190 kilometer–long fence, thereby restoring the most critical portion of the wildlife corridor. The government also has signaled willingness to remove more of the fence provided Namibia increases disease control measures.

— Courtesy of Gautham Rao, World Wildlife Fund (USA)

Uncertainty and Complexity of Corridors

Finally, biological corridors are highly uncertain and complex experiments. A current lack of empirical evidence demonstrating their benefits has not inhibited the planning or launching of hundreds of corridor projects worldwide. Yet most of these efforts are still in the planning stage and only a small proportion are under implementation. In addition to untested scientific issues, the design and especially the implementation of corridors involve a wide range of largely unexplored socioeconomic issues upon which long-term success depends. Given this uncertainty and complexity, building support for corridors from multiple stakeholders is especially challenging —

as is apparent in fledgling efforts like those along the floodplain of the Lower Kinabatangan River in Sabah, Malaysia, where stakeholders have highly conflicting interests and have yet to define a coherent corridor vision.

How Do You Build Support for Corridors?

As mentioned above, corridors are complex and long-term initiatives that require support from diverse sectors of society. Many of the cases presented in this book indicate that successful corridor initiatives have well-defined leadership, strong institutional coalitions, and broad public support. We examine each of these needs below.

Defining Leadership

First and foremost, it is essential to get visionary, motivated, and, if possible, strategically placed people committed and involved. Our case studies show that once such people are on board, other individuals and their institutions will follow. This is especially apparent in the case of the Veluwe, where the process of resolving entrenched conflicts between governmental institutions — which led ultimately to the formulation of an ambitious landscape corridor initiative — began through contact between two individuals placed in key governmental institutions at the provincial and national levels. Likewise, a few influential local people and scientists developed a compelling vision for the Y2Y Corridor, and another small group of scientists began efforts to reverse the extinction of the golden lion tamarin in Brazil's Atlantic Forest. Experience worldwide, however, shows that surprisingly few individuals play key roles in launching ambitious conservation initiatives.

Local leadership of corridor initiatives has three major advantages over initiatives run by outside actors or institutions. First, local leadership can design appropriate solutions based on knowledge of the local context. Second, local leadership reduces the perception of outside interference in local affairs. And third, local leadership is essential for motivating the local involvement needed to sustain corridors in the long run.

While local leadership is critical, nonlocal actors such as international nongovernmental organizations (NGOs) often play critical roles in large-scale, landscape corridor initiatives. One such role involves direct financing, which is provided by WWF in the Terai Arc and Atlantic Forest, and TNC in the Talamanca corridor. NGOs are also effective in facilitating or leveraging funding from other sources. TNC provides this supporting function in Florida, where it helps put landowners interested in conserving their land in touch with government agencies with funds for conservation. Technical assistance is another key role played by major NGOs, especially in developing countries. In the Talamanca corridor, for example, TNC has helped develop procedures for acquiring conservation easements.

It is important, however, that nonlocal NGOs do not take center stage in corridor implementation, and that local individuals, institutions, and communities remain at the forefront. An illustrative example of this is the Y2Y Corridor, where an umbrella organization (Y2Y) disseminates relevant research findings and information to local conservation efforts, provides small grants to support local implementation, and helps local organizations and communities publicize conservation efforts.

Involving Institutions and Coalitions

As mentioned above, a few visionary people often initiate corridor initiatives. While institutions usually follow later, these can develop into coalitions that play crucial roles in getting corridor initiatives off the ground. Such coalitions come from a relatively narrow segment of society committed to similar values, such as environmental NGOs, research centers, and other nonprofits. In Brazil's Atlantic Forest, for example, an international coalition comprising major zoos, research centers, and conservation groups — joined by their in-country counterparts — has: (i) nearly tripled wild populations of the golden lion tamarin, (ii) generated scientific knowledge critical for conservation, (iii) raised public awareness worldwide about this species, and (iv) begun building a corridor to link isolated populations. In Costa Rica, a coalition of conservation groups with expertise ranging from sea turtles to sustainable forestry helped launch the Talamanca corridor. In the Terai Arc, multiple nonprofit groups — especially in Nepal — are collaborating actively in corridor design and implementation.

Over time, broad coalitions consisting of public, nonprofit, and private sector institutions can provide a powerful impetus for corridors. Along the Kinabatangan River, WWF Malaysia has begun to win the trust of diverse local groups such as tourism companies, oil palm estate owners, riverine communities, and relevant governmental agencies — thereby providing a basis for embarking on a corridor initiative. In this case it would have been impossible, and probably counterproductive, to organize a coalition of multiple institutions at the outset. Likewise, local government agencies have taken the lead in building diverse coalitions in support of corridor initiatives in the Y2Y region of Canada and in the Veluwe region of the Netherlands.

Broadening Public Support

Whether through the efforts of visionary individuals, a single institution, or an institutional coalition, the next step in a corridor initiative is to broaden public support and, at the same time, reduce current or potential resistance. This in turn requires knowledge of both potential allies and adversaries of the initiative, which can be achieved through a stakeholder analysis. Such an analysis should be straightforward and designed to indicate:

Who are the key interest groups?
What are their interests in relation to environmental issues in general, and to the corridor specifically?
How could they contribute to or undermine corridor efforts?
What are appropriate strategies for mobilizing them (in the case of key institutions or coalitions), winning them over (converts), or reducing their opposition (adversaries)?

Broadening support for and overcoming opposition to corridor initiatives requires knowledge about those segments of society likely to have an impact (either favorable or unfavorable) on public opinion — such as NGOs, the scientific community, the media, relevant governmental agencies, businesses, and policymakers. Groups such as landowners that directly control areas within the corridor (or areas that generate significant impacts on it) are critical, and if possible, their interests and attitudes should receive consideration. For example, in the Atlantic Forest, long-term environmental education efforts by NGOs have transformed local attitudes toward the golden lion tamarin. Today, formerly hostile or indifferent landowners vie to replenish local tamarin populations.

Building support for biological corridors often requires linking these initiatives to critical environmental services on which people depend. Threats to water-related services are especially effective in mobilizing support. Perceived declines in water-related services — such as water quality, modulation of flooding regimes, and maintenance of fisheries — are building constituencies for conservation in the Lower Kinabatangan, the Atlantic Forest, the

Terai Arc, Panther Glades, and the Cascade–Siskiyou National Monument. Threats to services that sustain local economies can be especially effective in motivating action. In the Muskwa–Kechika Management Area of the Y2Y Corridor, for example, growing concern among hunters and backcountry guides about loss of wildlife due to extractive industries is helping to build local support for zoning and other conservation measures.

Threats to cultural values and lifestyles provide additional motivation for supporting corridors. In the Y2Y Corridor, growing concern over rapid and uncontrolled development motivated residents of Canmore, Canada, to designate a corridor and develop a growth-management plan. Similarly, increased road building, industrialization, and unrestricted tourism have helped build local support for the Muskwa–Kechika Management Area.

While it is critical to broaden support for corridors, it is also important to recognize that some groups will remain opposed, and in such cases the best strategy may be to isolate them or neutralize their opposition. In the Veluwe region, for example, the national planning ministry was strongly opposed to greater collaboration among government agencies, but other ministries — which had been won over to an ambitious corridor vision for the region — overruled its opposition. In Klamath–Siskiyou, extractive industries have led a systematic campaign pitting jobs against environment. Here efforts to change the ingrained attitudes of those industries may be less productive than attempting to influence broader perceptions through the media. In this context, an excellent strategy to build public interest in and support for corridors is to make them tangible and relevant to people's lives. A hike along the Y2Y Corridor by Karsten Heuer of Parks Canada, during which he gave presentations in communities along the way, greatly increased public awareness of the corridor and its potential benefits. Elsewhere in the Y2Y Corridor, backcountry guides, to build support for conservation, took political leaders and reporters on tours of the Muskwa–Kechika Management Area.

Likewise, gaining and maintaining public interest requires achieving early successes and building on them. Thus it is critical not to attempt to build a corridor all at once, but instead to begin with highly visible segments where the probability of success is high. In the Atlantic Forest corridor, for example, initial efforts are focusing on a few properties belonging to highly committed landowners dispersed along the corridor length. Instead of restoring forest cover through natural regeneration, corridor proponents will establish more costly plantations of native species to generate quick results and thereby show landowners that building a corridor is possible.

A key issue involves convening or consulting with interest groups — and specifically whether, when, and how to do so. Involving diverse interest groups appears to be essential for establishing and maintaining corridors. It is a common belief that in environmental initiatives involvement of all relevant groups should be encouraged early on (Hay 1990, Gailus 2000). Our cases show that, depending on local conditions, a variety of strategies for involving such groups may be more appropriate. Thus, planning for a Kinabatangan River corridor is under way, but it is still too soon to bring together all major groups to discuss this idea. At the other extreme, since the outset of corridor efforts in the Veluwe region, a highly diverse spectrum of stakeholders has been involved in design and implementation. In the Terai Arc, Y2Y, and Klamath–Siskiyou, active or potential corridor supporters were consulted initially and many were convened to participate in corridor planning. Such selective involvement is important to form coalitions (if appropriate), define objectives, and develop strategies to attain those objectives — which would be impossible to accomplish in a more inclusive but confrontational environment. Once the institutional roles, objectives, and strategies for establishing a corridor have been determined, convening a fuller range of interest groups — including adversaries — makes more sense.

While essential, involving diverse interest groups is no guarantee of success (box 4.2). It has worked very well in the Veluwe region and in the Bow Valley in Canmore, Canada (Y2Y). In the Talamanca case, a series of meetings between environmental NGOs and landowners has helped overcome the latter's initial resistance to a perceived loss of resource-use rights caused by the proposed corridor. Yet it is still too early to assess whether involving diverse interest groups has been positive in the Muskwa–Kechika region of Y2Y, and the results of convening community representatives and government agencies in the Terai Arc have been mixed. As mentioned above, in the Kinabatangan River region convening diverse interest groups or consulting them about a corridor was until recently considered potentially counterproductive.

BOX 4.2. BUILDING CONSENSUS AMONG LOCAL INTEREST GROUPS:
TWO CONTRASTING EXPERIENCES

Within the Y2Y Corridor, the Muskwa–Kechika Management Area (MKMA) of British Columbia, Canada, provides an example of effective consensus-based planning and management assisted by local government. During two years of negotiation, local environmentalists and representatives from forestry, mining,

continued

and petroleum industries crafted a plan for multiple-use management in this area. The provincial government provided a neutral forum in which the various interest groups could meet, and it subsequently established a trust fund to pay for implementing the management plan. An advisory board composed of representatives from diverse interest groups is responsible for monitoring enforcement of the management plan. The long-term success of MKMA remains to be seen, but much progress has been made because most local stakeholders respect the Muskwa–Kechika plan as the product of a fair and open consensus process in which their interests were represented.

Consensus-based planning and management have been less successful in the case of Cascade–Siskiyou National Monument of Oregon, USA. Support from local landowners is particularly crucial for successful implementation of this corridor because much of the proposed monument land is privately owned (38 percent). The 1999 designation of the monument was preceded by public hearings over fifteen years sponsored by local environmental organizations and the Federal Bureau of Land Management. However, strong resistance by well-organized timber and ranching interests has delayed the defining of a management plan for the monument. These groups view the monument as imposing new restrictions on the BLM-held public lands they previously utilized. In this case, governance through local consensus has broken down.

— Source: Cases 4 and 5, respectively

In short, key strategies for building support for corridors include:

- getting a few visionary, motivated, and, if possible, strategically placed people on board;
- developing links with institutions committed to similar values, such as environmental NGOs, research centers, and other nonprofits, to form coalitions (if appropriate), define objectives, and develop strategies to attain those objectives; and
- carrying out stakeholder analysis to determine who the main actors are, what their interests are, how they could contribute to (or undermine) corridor efforts, and what the best strategies are for getting them involved or overcoming their opposition.

What Are Effective Incentives for Corridors?

Except in highly specific cases, command-and-control approaches generally are not appropriate for implementing biological corridors. Such approaches are unlikely to succeed over the long term. Corridor implementation requires many more carrots than sticks — especially in the case of large-scale, landscape corridors. Economic incentives have increasingly been found to be far more effective in motivating people to support and become involved in corridor implementation. Both negative and positive incentives can be effective. Foremost among the negative incentives are loss of biodiversity and loss of critical environmental services, which is generating growing concern in many cases. For example:

- In the Atlantic Forest, increased environmental awareness of the threats facing the golden lion tamarin has motivated landowners to establish private reserves.
- In the Muskwa–Kechika Management Area in the Y2Y Corridor, concern among hunters and backcountry guides about loss of wildlife due to extractive industries is leading to increased local support for zoning and other conservation measures.
- Local concern over declining water-related services — such as water quality, modulation of flooding regimes, and maintenance of fisheries — is building constituencies for conservation in the Lower Kinabatangan, the Atlantic Forest, the Terai Arc, Panther Glades, and the Cascade–Siskiyou National Monument.

Threats to cultural values and lifestyles provide additional incentives for supporting corridors. In the Y2Y Corridor, growing concern over rapid and uncontrolled development motivated residents of Canmore, Canada, to designate a corridor and develop a growth management plan. Similarly, increased road building, industrialization, and unrestricted tourism have helped build local support for the Muskwa–Kechika Management Area.

Finally, biological corridors provide potentially strategic spaces for defining and, eventually, enforcing resource-use restrictions through negative incentives such as fines. As corridor implementation proceeds and public support grows, one would expect that enforcement should become increasingly important. Yet our cases reveal few examples of enforcement. Strict zoning arrangements have been imposed in the Muskwa–Kechika Management Area to ensure long-term ecosystem integrity. In the Veluwe region, planners have proposed reducing speed limits on secondary roads as a way to minimize environmental impacts, but they have yet to implement these limits. Burning pastures and cropped areas is technically illegal in the Atlantic Forest, although this policy is rarely enforced. It is likely that enforcement will be more successful in smaller-scale initiatives such as linear corridors, and in places with well-developed judicial systems.

Because they usually extend over areas owned by diverse groups, corridors almost inevitably require positive as well as negative incentives. A variety of policies providing positive economic incentives for conservation have emerged in recent years and are applied actively worldwide. Generally designed to encourage conservation of environmental services such as watershed protection, these incentives can be organized along a public-to-private gradient that includes: (i) direct payment schemes run by public agencies, (ii) public–private trading schemes, and (iii) self-organized deals between private entities (Johnson et al. 2001). Such incentives play an increasingly important role in both local and large-scale conservation initiatives, and as a result, they are likely to provide a strategic tool for corridor implementation at all scales. (See a capsule description of each type and illustrative examples in box 4.3.)

BOX 4.3. ECONOMIC INCENTIVES FOR CONSERVATION

Direct payment scheme (public). Here a public agency pays a public or private entity to maintain or restore an environmental service. In contrast to regulatory measures that often lead to preemptive resource degradation as a way to avoid compliance, such payments create a direct financial incentive for conservation on the part of the service providers (Ferraro and Simpson 2002; Ferraro and Kiss 2002, 2003). This type of publicly driven incentive is the most widespread today, and it is likely to remain so because environmental services usually transcend the interests of specific groups and provide benefits for a broad public (Johnson et al. 2001).

For example, in the city of Quito, Ecuador, 1 percent of the revenue from hydropower generation and water-use fees goes into a fund designated specifically for protection of the Cayambe–Coca and Antisana reserves, which are important upstream sources of the Quito water supply. The fund revenues go primarily to pay property owners near the reserves to change their land-use practices. The next step is to base the fee on a more complete valuation of the environmental services that the reserves provide (Chomitz et al. 1998).

Trading schemes (public–private). Trading schemes are emerging in countries with tightly regulated environmental standards such as strictly controlled water consumption or pollution limits (or "caps"). The government provides permits for the use of a resource and sets the caps. Yet it does not determine how to meet the caps, thereby creating markets for trading permits between parties that are under the cap and those that exceed it. An analysis of the advantages and disadvantages of different marketable permits suggests that, in general, they work

---

continued

best at larger scales and to control point sources of pollution (Rose 2000). A strong regulatory framework and effective enforcement are key requirements for trading schemes (Johnson et al. 2001).

In the midwestern United States, for example, tightly regulated, so-called point sources of pollution such as factories are paying unregulated, "nonpoint" sources such as farmers to reduce their emissions — presumably at a far lower cost than if the factories were to reduce their own pollution. In Australia, land clearing has increased salinization problems for irrigators in the Murray–Darling Basin. Here much of the vegetation that once transferred water from the ground to the atmosphere (through evapotranspiration) is now gone, and as a result water tables have risen, bringing dissolved salts to the surface. A market scheme is now being piloted by which landowners who plant trees receive credits that may then be purchased by farmers who benefit from the increased availability of water suitable for irrigation (Perrot-Maître 2000).

Self-organized deals (private). Private entities are making deals to pay for environmental services in river basins, with little or no government involvement. These cases are more likely to occur where there is no effective regulatory system in place or where a private interest needs water quality or flow that is above regulatory standards. Self-organized private deals involve low transaction costs, which suggests an important benefit in comparison to schemes involving greater public intervention.

The French company Perrier-Vittel, for example, is the world's largest bottler of natural mineral water. Its most important water sources are in heavily farmed watersheds, where nutrient runoff and pesticides threaten the aquifers upon which the company relies. Instead of building more expensive filtration plants, Perrier-Vittel has opted to reforest sensitive infiltration zones and finance farmers who switch to organic agricultural practices (Perrot-Maître 2000).

The Nature Conservancy has long been involved in purchasing property and conservation easements to protect critical conservation areas in the United States. It is now expanding use of those instruments to Latin America (Randall Curtis, personal communication).

Below, we briefly review some of the major positive incentives presented in the case studies.

Compensation to Property Owners

Corridor implementation arrangements can involve compensation to property owners through mechanisms such as purchasing land or easements

to land prioritized for conservation, direct payments for environmental services or for conservation costs, and tax breaks for property owners who set
aside their land for conservation. For example:

- In both Panther Glades and Pinhook Swamp, the state of Florida provides funds for purchasing property or conservation easements.
- In the Atlantic Forest, areas allocated for private reserves are exempt from
  land taxes.
- In both Y2Y and Cascade–Siskiyou, new policies are being proposed to
  provide tax breaks for conservation easements.
- In the Klamath–Siskiyou ecoregion, nonprofits such as WWF help small
  proprietors cover the costs of timber certification using standards established by the Forest Stewardship Council.
- In Costa Rica, a national program provides direct payment to landowners
  who protect water supplies and ecosystems critical for biodiversity conservation; plans are also under way to provide compensation that will
  ensure that natural forests maintain their critical role as sinks for carbon
  (Chomitz et al. 1998).
- In Brazil, new legislation authorizes the establishment of watershed committees to determine appropriate water-pricing policies and priorities for
  investing the additional revenues in water-related services. These revenues could accrue to property owners who protect or restore watersheds
  (Porto et al. 1999).

Increased Public Revenues

Incentives for conservation initiatives — some of which involve biological corridors — are beginning to flow to governments at various levels. For
example:

- In the Atlantic Forest, municipalities that conserve a larger portion of
  their areas in protected areas receive a greater share of tax revenues from
  Brazilian states.
- Local governments are receiving substantial revenues from tourism, especially in the Veluwe.
- The state of Florida provides rural counties in the Panther Glades corridor with payments to compensate property tax revenue lost when private lands are purchased for conservation.

- In Panther Glades and Pinhook Swamp, water-management districts have contributed substantial funding to a new state-funded program ("Florida Forever") to purchase lands or easements that protect watersheds for downstream ecosystems and nearby cities in addition to panther habitat.

Tourism

Estimates over the past decade show that the tourism industry is one of the fastest-growing economic sectors worldwide, and ecotourism is thought to be the fastest-growing part of this industry. Examples of the increasing importance of tourism in corridors follow.

- In Panther Glades, tourism via a state-funded "rails-to-trails" program was important for winning Hendry County's support for the corridor.
- Tourism in the Veluwe region generates gross receipts of US$1 billion per year and provides critical justification for corridor implementation.
- In the Atlantic Forest, the potential for increased income from ecotourism provides a major incentive for establishing private reserves.
- Ecotourism is a fast-growing business in the Lower Kinabatangan River and supports forest protection initiatives there.
- In 2000, tourism in and around the two most popular national parks in the Terai Arc, Royal Chitwan and Royal Bardia, generated about US$1 million in gross revenues. Local communities in Nepal received 30–50 percent of this revenue, and they raise additional income by charging fees to trekkers crossing outside the national parks.

Technical Assistance

Most corridor initiatives provide diverse forms of technical assistance to local stakeholders, and this can provide yet another incentive for corridor implementation. For example:

- In Florida, TNC provides scientific expertise, pays for surveys and assessments of potential conservation land for state purchase, trains state agencies in easement monitoring, and helps landowners develop easement plans.
- In the Atlantic Forest, corridor initiatives provide environmental education (in the case of the golden lion tamarin) and agricultural extension (in the case of the black lion tamarin — see box 4.4).

- Nonprofits in the Terai Arc region help rural people gain access to governmental services such as family planning, provide childcare to help women get more involved in resource management, supply seedlings for reforestation, and give grants for small-scale ecotourism.
- In the Talamanca corridor, nonprofits help small landowners register to receive ecosystem service payments from the Costa Rican government.

BOX 4.4. INCENTIVES FOR BUILDING A CORRIDOR FOR THE BLACK
LION TAMARIN IN BRAZIL'S ATLANTIC FOREST

The black lion tamarin was considered extinct until its rediscovery in 1970 in southwestern São Paulo state, Brazil, where it is now protected in the 36,000-hectare Morro do Diabo state park. Outside the park are a number of forest fragments, including a large, 2,000-hectare upland forest block connected to the park by a 4 kilometer–long strip of riverine forest, which provides a corridor used by the black lion tamarin and other vulnerable species such as tapirs (*Tapirus terrestris*), ocelots (*Leopardus pardalis*), and pumas (*Puma concolor*). Combined, the natural corridor and forest block increase the area of the park by 5 percent. Yet this linear corridor has come under increasing threat from shifting cultivation and ranching practiced by formerly landless families who settled in the region in the 1990s.

To explore possibilities for implementing land-use alternatives, in 1996 a local NGO (IPÊ — Institute for Ecological Research) contacted local farmer communities and representatives of Brazil's Movement for Landless People (MST). The farmer communities were interested in alternatives that could complement family income, and the MST wanted to change its "anti-ecological" image in Brazil. These contacts resulted in a project to establish an agroforestry buffer along the margins of the forest corridor, thereby diminishing the edge effect caused when cleared areas are located adjacent to forest. Begun in 1998, the project provides short courses that illustrate the various benefits provided by agroforestry — including increased soil fertility and decreased fertilizer requirements, need for weed and pest control, and erosion. IPÊ and community members have established nurseries that together produce about 100,000 seedlings per year, representing species that produce timber, fruits, fuel, fodder, and shade for livestock, as well as environmental services such as nutrient cycling and attraction of birds and other dispersal agents. With the assistance of IPÊ technicians and community extension agents hired by the project, families use the seedlings according to their needs but must plant at least 60 percent in the buffer zone adjacent to the forest corridor.

continued

Today, forty-five farmers have adhered to the project and are implementing agroforestry in the area adjacent to the corridor. By enabling local farmers to establish alternatives to predominant land uses and providing a protective corridor buffer, this strategy illustrates a promising option for conserving the region's remaining biodiversity.

— Source: Cláudio Pádua, personal communication

In short, a wide range of potential incentives exists to support corridor initiatives. In the Terai Arc, for example, the primary goal of conservation practitioners is to provide habitat connectivity for tigers, rhinoceros, and elephants. Multiple uses are necessary, however, because millions of people inhabit the region and many rely on natural resources for subsistence. Local communities benefit from restoration of degraded linkage areas through protection of water resources, income from ecotourism, and sustainable harvest of subsistence resources. Increasing the usable habitat for wildlife and natural barriers along the corridor route can also reduce human-wildlife conflict.

Other types of incentives are characteristic of a developed-country context. For example, incentives used to build corridors in Florida include tax breaks for private landowners who sell or place conservation easements, state government funds for purchase of corridor land and easements, compensation by the state government to rural counties for property tax revenue lost when private lands are purchased for conservation, and increased municipal tax revenues coming from recreation and other economic activities linked to conservation. The key when including multiple incentives for biological corridors is to ensure that additional uses do not compromise the primary objective of protecting biodiversity.

How Should Corridors Be Governed?

Because they may involve multiple sectors and cover extensive areas, biological corridors frequently require complex governance arrangements. Lack of clarity about such arrangements can lead to conflict — especially in the case of large-scale, landscape corridors. For example, in the early phases of the Veluwe corridor program, the national and provincial governments had constant disagreements, and there was little or no coordination among the six national agencies involved in the program. In Florida, some counties initially objected to allocating land to the Panther Glades corridor for fear of losing tax revenue from future development, and there has been long-standing disagreement among public agencies concerning the jurisdiction of lands designated for conservation.

Resolution of such conflicts usually requires convening relevant stakeholders, getting them to negotiate, and, if possible, establishing arrangements to possibly avoid future conflicts. For example:

- In the Veluwe region, a workshop convening major interest groups helped resolve conflicts among governmental agencies, and a formal governing body has been established to address future conflicts.
- In Talamanca, a corridor commission composed of diverse stakeholders makes management recommendations, monitors conservation easements, and reports violations of environmental policies to the national government.
- In the Cascade–Siskiyou National Monument, the Federal Bureau of Land Management will retain overall management responsibility, but the state of Oregon will handle specific responsibilities, such as issuing of hunting and fishing licenses.

A fundamental requirement for governance of corridors is *co-management*, in which the roles and responsibilities of relevant governing bodies (both governmental and nongovernmental) are clearly defined at all levels. Co-management is relevant when resource management issues are beyond the control of any single entity, thus requiring collaboration between them — which is the case for practically all corridors at a landscape scale. Such collaboration can range from an informal division of labor to a more formal definition of roles and responsibilities. Just as local jurisdictions lack control over actions outside their boundaries, national governments lack sufficient knowledge of local conditions as well as the manpower needed for monitoring and enforcement, and nongovernmental organizations are only partially representative of stakeholder interests because they promote specific interests.

Various cases illustrate the concept of co-management.

- In the Muskwa–Kechika Management Area in the Y2Y Corridor, the Canadian government has helped set up so-called planning tables to involve communities in land-use planning. Tables use a consensus process to develop plans and oversee implementation, while the provincial government provides funding and approves plans. Funds for implementation come from a trust fund established by the government that matches contributions from other (mostly local) sources and is overseen by the community-based advisory board.

- In the Terai Arc, local "community forest user groups" develop management plans for government land leased to them, while the national government approves the plans and monitors their implementation.
- In the Talamanca corridor, a governing commission — composed of local representatives of environmental groups, landowners, and other stakeholder — makes management recommendations and monitors conservation easements, but has no power of enforcement and must report any violations to the Costa Rican government.

Of the cases reviewed in this book, the Veluwe program provides the best example of co-management, with distinct yet overlapping roles and responsibilities exercised by an extremely wide range of stakeholders (box 4.5).

BOX 4.5. CO-MANAGEMENT OF THE VELUWE CORRIDOR

Co-management arrangements for designing and implementing the Veluwe corridor involve institutional actors at a variety of levels. At international and national levels, both the European Union and the Dutch government have established a broad policy framework encouraging landscape linkage at diverse levels. The expectation is that the Dutch government will provide a substantial part of the program's budget during 2001–2010, and additional support is anticipated from the European Union. National funds will be channeled through six ministries, each of which may establish specific targets but (ideally) will not determine how to reach those targets. National government representatives also participate in the governing body that sets broad program objectives and strategies.

The provincial government coordinates the entire program, provides partial funding, and leads both the governing body and a working group that assists in developing specific projects. The province has played a pivotal role in conceiving the program, continues to encourage broad stakeholder involvement, and provides the necessary oversight to ensure that program directions are followed and strategies carried out. It prepares periodic reports on overall program performance, which it distributes to funding agencies and which are also available for consultation by all program participants and interested citizens. Provincial government agencies are also involved in carrying out specific projects under the program.

continued

Finally, a diverse array of local interest groups is involved in developing and executing projects supported by the program, including provincial government agencies, townships, nonprofits, private landowners, and a local business association. Projects range from improving tourist facilities to landscape restoration. Local actors are also represented on the program's governing body and working group.

The Veluwe program reveals several principles that are highly relevant to corridor governance:

• First, there is a clear hierarchical division of roles and responsibilities. National and international government agencies furnish most of the funding and set broad policy directions and more specific targets. The provincial government coordinates all aspects of the program, provides oversight, and reports to funders. Local actors execute the program through specific projects.

• Second, the program's governance arrangements provide a forum for all participants to interact. By setting broad objectives and strategies, the governing body helps to keep the program from evolving into a mere collection of projects. More technically oriented representatives participate in the working group, which reviews and approves projects.

• Third, by limiting participation to fifteen members, the program's governing body attempts to strike a balance between representation and decision-making capacity.

While co-management appears to be a promising option for governance of corridors, it is extremely complex in practice and probably works best in countries with strong democratic traditions, where collaboration between governmental agencies at various levels and civil society is encouraged. In many countries, conditions may not be propitious for applying co-management as a governance option for corridors.

Finally, as large-scale conservation initiatives such as landscape corridors cross national frontiers, the complexity of governance arrangements will increase and disputes over rights and resources are likely to become more common. For the Terai Arc corridor, for example, in India state agencies carry out most of the conservation work, whereas in Nepal NGOs and community organizations play this role. These differences could thwart attempts to develop coordinated conservation strategies and actions across the corridor — such as the development of antipoaching regulations and land-use planning.

One alternative for addressing governance conflicts at the international level, while also conserving large-scale ecological phenomena such as migrations, is so-called transfrontier conservation areas (TFCAs). TFCAs were originally conceived of as protected areas straddling frontiers between two or more countries that are managed jointly by those countries, with a view to increasing the effectiveness of conservation. The concept has since expanded, and today TCFAs extend far beyond national parks and game reserves, incorporating private land, communal land, forest reserves, and areas managed for consumptive use of wildlife (Hanks 2003). The first such area, established by Canada and the United States in 1932, encompasses the Waterton and Glacier National Parks in an area that is now nested within the Y2Y Biological Corridor. Today at least 169 TFCAs involve 113 countries and cover over one million square kilometers (van der Linde et al. 2001).

Due to the critical ecological and economic importance of migrating wildlife, a high concentration of TFCAs — known regionally as "peace parks" — have been established or proposed in southern and east Africa. The first one unified the Gemsbok National Park in Botswana and the Kalahari National Park in South Africa, which since 1948 has been managed as a single unit to permit free movement of wildlife and was formally declared as the Kgalagadi Transfrontier Park in 2000. One TFCA, advocated by Conservation International, would connect protected areas in Botswana, Namibia, Zambia, and Zimbabwe (Conservation International 2001; fig. 4.2). If implemented, this TFCA will represent a landscape corridor immense in both scale and value for biodiversity conservation.

The new TFCAs established or planned worldwide provide opportunities for learning about how to govern large conservation landscapes spanning two or more countries. Yet because TFCAs consist primarily of public or communal areas designated for some form of protection, they avoid many of the challenges faced by landscape corridors that include extensive privately owned areas.

An Integrated Implementation Strategy

The specific issues discussed above — such as resource management, incentives, and governance — should be part of an integrated corridor implementation strategy. This strategy requires diverse approaches that are likely to vary from place to place. Yet several general approaches can be predicted that vary according to scale.

In linear corridors, the primary objectives are to protect target species by facilitating their movements, and to maintain or restore linkage of habitat fragments or local ecosystem services. Depending on the habitat quality of

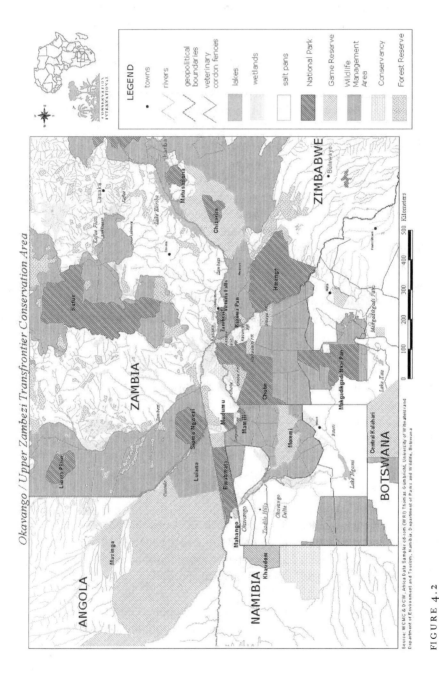

Okavango / Upper Zambezi Transfrontier Conservation Area

LEGEND

- towns
- rivers
- geopolitical boundaries
- veterinary cordon fences
- lakes
- wetlands
- salt pans
- National Park
- Game Reserve
- Wildlife Management Area
- Conservancy
- Forest Reserve

ANGOLA

ZAMBIA

NAMIBIA

BOTSWANA

ZIMBABWE

0   100   200   300   400   500   Kilometers

Source: WCMC & DCW, Africa Data Sampler cd-rom (World); Thomas Gumbricht, University of Witwatersrand and Department of Environment and Tourism, Namibia; Department of Parks and Wildlife, Botswana

FIGURE 4.2

One TFCA advocated by Conservation International would include Hwange and Zambezi National Parks (Zimbabwe); Mamili, Mudumu, and Bwabwata Parks in the Caprivi region (Namibia); the Sioma Ngwezi and Kafue National Parks (Zambia); and Chobe and Makgadikgadi Nxai Pan National Parks and the Moremi Game Reserve (Botswana). Source: Conservation International 2001.

the corridor, these objectives may be achievable over a relatively short term, and they usually involve a limited set of strategies, such as purchase of land or easements or enforcement of strict zoning requirements — which may be most likely to succeed at smaller scales. For example, one of the primary conservation objectives in the Veluwe region is to restore movement by red deer and wild boar. This is accomplished by constructing ecoducts that enable these and other species to cross major highways. At smaller scales, it also is critical to identify objectives achievable over a short term — such as restoring a strategic part of a linear corridor — to sustain the interest of local landowners and attract new support. This approach is being utilized in building a linear corridor to facilitate movement of the golden lion tamarin in the Atlantic Forest.

In contrast, landscape corridors generally involve more numerous and broader objectives, such as biodiversity protection, conservation of ecological processes, sustainable use of natural resources, and recreation. Achieving or maintaining such diverse objectives requires a wide range of tools — including land-use zoning, establishing public and private protected areas, and using incentives for environmentally benign forms of resource and habitat use. For example, policies providing incentives for easements or forest-based services can promote conservation across large-scale landscape corridors. Such incentives are on the horizon in the Y2Y Corridor and in the Atlantic Forest of Brazil, and they already are functioning in Talamanca and the Terai Arc. The landscape approach is more or less a form of regional planning designed to promote a mix of resource and habitat uses. Because landscape corridors can incorporate linear corridors, implementation may involve combining top-down actions such as enabling policies with bottom-up actions such as habitat restoration or protection in strategic locales.

These concepts provide a general framework for corridor design and implementation. With hundreds of corridor initiatives under way worldwide, yet most in incipient stages, there is a unique opportunity now to define such a framework and a uniform set of terms that will facilitate understanding and exchange of experiences and lessons in the future. The unifying theme that underlies this framework is connectivity, whether at extremely small or at immense scales. While further scientific research is needed to understand the various functions that connectivity provides, most conservation biologists and practitioners agree that this is the only viable way to combat the devastating process of fragmentation — which is threatening species, habitats, ecosystems, and ecological processes throughout the globe.

This chapter has reviewed major issues that resource-use managers are likely to confront in implementing biological corridors. Because corridors are a relatively new concept, much of the information we have presented comes from experiences that are still in their beginning stages. It is clear, however, that corridors — like other ambitious conservation initiatives — are complex undertakings that transcend strictly technical issues and involve a variety of other matters, such as stakeholder engagement and governance, that are difficult if not impossible to predict or control. While enthusiasm currently is high, this and previous chapters emphasize the fact that corridors are not silver bullets. The success of corridors as a conservation strategy will require long-term commitment and a willingness to tackle complex issues.

5 Case Studies

Methodology

Upon embarking on the search for corridor cases for this book, we expected that it would be possible to select only corridors under full implementation. Yet despite the existence of probably hundreds of corridor initiatives worldwide,[1] and after examining dozens of cases, we found that almost all are in the planning stage, extremely few are under implementation, and none has been under way for a sufficient time to allow a full evaluation of its results. This discovery led to a revision of the criteria for selecting potential case studies for this book. While implementation was still a factor in selection, we identified approximately twenty-five cases of potential interest that exhibited a wide geographic distribution and variety of habitats. In addition, each case offered potentially interesting insights regarding corridor design and implementation, and had specific sources of information.

Obtaining information from these sources became a critical limiting factor in assembling the case studies. As a result, of the twenty-five cases initially identified, a total of eight are presented here (fig. 5.1):

Case 1: Atlantic Forest Corridor, Brazil
Case 2: Talamanca–Caribbean Corridor, Costa Rica
Case 3: Pinhook and Panther Glade Corridors, Florida, USA
Case 4: Yellowstone to Yukon (Y2Y) Corridor, USA and Canada
Case 5: Klamath–Siskiyou Corridor, USA
Case 6: Lower Kinabatangan River Corridor, Sabah, Malaysia

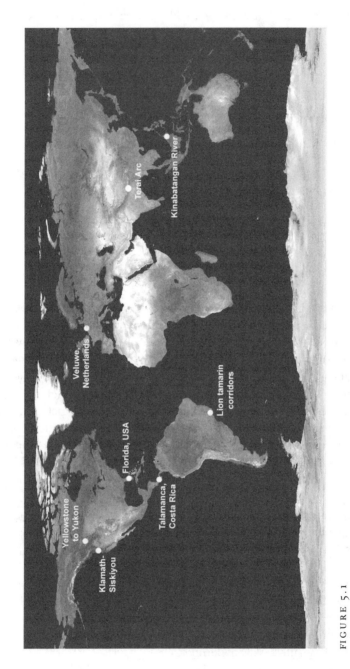

FIGURE 5.1

Locations of corridor case studies.

Case 1: Forest Corridors for Lion Tamarins in the Atlantic Forest

Summary

This case involves efforts to design and implement corridors to protect the highly threatened lion tamarins in Brazil's Atlantic Forest region.[2] It focuses on the golden lion tamarin (*Leontopithecus rosalia*) in Rio de Janeiro state but also provides information about the black lion tamarin. The major issues in this case are described below.

Design and Management   A 20 kilometer–long habitat corridor is designed to link fourteen isolated fragments of golden lion tamarin habitat, crossing the land of numerous private landowners. Long-term studies indicate that a minimum 30 meter–wide corridor would permit animal movement between fragments, but a considerably greater width would be necessary to establish permanent linkage in a fire-prone landscape. Establishment and long-term maintenance of the corridor requires effective protection against cattle grazing (in the establishment phase) and burning, and the corridor could provide a potential conduit for fire.

Economic and other incentives and stakeholder engagement can help to improve project acceptance and overcome any reluctance to engage in these extra management requirements.

Stakeholder Engagement   Diverse institutions actively collaborate in conserving the golden lion tamarin, including international donors, conservation organizations, and research centers, together with corresponding institutions in Brazil. This collaboration, coordinated today by a local environmental NGO, the Golden Lion Tamarin Association (AMLD), has been critical in reversing the downward population trends, increasing public awareness, and raising substantial funds for conservation efforts. Long-term environmental education efforts have transformed the attitudes of landowners toward the golden lion tamarin. Today, they are generally willing to collaborate with environmentalists and scientists in seeking appropriate management practices for the golden lion tamarin habitat. Implementation of this project is focusing on a few strategic locales along the corridor where early success can be achieved, thereby providing a demonstration effect for other landowners.

Economic and Other Incentives   The institutions currently collaborating in conserving the golden lion tamarin, in particular the donor agencies but also a major Brazilian forestry company, are supporting efforts to establish the corridor, and AMLD will play a major role in its implementation. Increasingly, local landowners are taking an interest in the tamarin for both aesthetic and, to a lesser degree, economic reasons (such as potential support for ecotourism), and five have committed to establishing the corridor on their lands. New water legislation in Brazil opens the possibility of using higher water tariffs to support forest conservation and restoration, which would provide a powerful economic incentive for this and other corridor initiatives.

Background

Brazil's Atlantic Forest region is generally considered one of the top hotspots for terrestrial biodiversity on earth. It once covered over 1 million square kilometers but has now been reduced to less than 10 percent of this area, primarily because of expanding agricultural and industrial activities. The region contains large metropolitan areas, including Rio de Janeiro and São Paulo.

The Atlantic Forest claims 849 species of birds, 188 of which are endemic — including the white-necked hawk, the purple-winged ground dove, and the blue-cheeked parakeet.[3] The region contains 20,000 species of plants, roughly half of which are endemic. Over 400 tree species (a world record) were reported recently from a single hectare in southern Bahia state. In terms of birds and plants, the diversity in the Atlantic Forest rivals that of the entire United States. The region has a high concentration of endangered species, including the colorful lion tamarins, which have become flagship species for ongoing efforts to restore the Atlantic Forest.

Endemic to the Atlantic Forest, all four species of lion tamarins — including the golden lion tamarin (*Leontopithecus rosalia*), the golden-headed lion tamarin (*L. chrysomelas*), the black lion tamarin (*L. chrysopygus*), and the black-faced lion tamarin (*L. caissara*, discovered only in 1990) — are threatened by forest extraction, ranching, and urban expansion, which lead to habitat fragmentation. This case focuses on the golden lion tamarin, a strikingly beautiful squirrel-sized primate that has been the target of unprecedented conservation efforts in Brazil and internationally since the early 1970s. Formerly distributed throughout the lowland coastal forests from southern Espírito Santo state to southern Rio de Janeiro, today this species is restricted to small forest fragments in seven municipalities in the São João river basin in central Rio de Janeiro state. Established by the Brazilian government in 1974, the Poço das Antas Biological Reserve represents the largest forest fragment (6,300 hectares) and contains the largest tamarin population (estimated at 220 individuals in 2000).

Key elements of current conservation efforts to restore the golden lion tamarin involve *ex situ* breeding in zoos, reintroduction into the wild, and translocation of small, isolated groups into larger forest fragments, where they have greater probability of survival. From 1984 to 2000, 156 captive-born tamarins were released into the wild, resulting in a net increase of 359 animals due to subsequent reproduction. These efforts have enabled wild populations of the golden lion tamarin to increase from an estimated low of under 200 individuals in the late 1960s to about 1,000 in 2001. In addition, translocation of groups inhabiting small, isolated fragments into a larger forest area previously without tamarins led to new populations. For example, the federal government established the 3,200-hectare União Biological Reserve, now containing an estimated population of 120 individuals. Today the area dedicated to golden lion tamarin protection — in both the two public reserves and numerous privately owned forests — totals 16,600 hectares.

The next milestone, targeted to occur by 2025, is the protection of 2,000 animals in 25,000 hectares of Atlantic Forest habitat.[4] Because of the high fragmentation of forests in the region, achieving this goal will require establishing corridors of native vegetation between currently isolated forest fragments, thereby enhancing breeding opportunities and improving the genetic health of tamarin populations.

Design and Implementation

The Golden Lion Tamarin Association (AMLD) has assumed responsibility for the design and implementation of the landscape corridor. Since its founding in 1992, AMLD has coordinated golden lion tamarin research and conservation activities. Initially, AMLD has defined thirteen small linear corridors that would link fourteen isolated fragments of tamarin habitat.

Together these smaller corridors eventually would coalesce to form a single, larger corridor approximately 20 kilometers in length, connecting the Poço das Antas Biological Reserve to the north with a 1,200-hectare forest block on the Rio Vermelho Farm to the south (fig. 5.2). This block represents the largest privately held forest area containing a golden lion tamarin population, comprising approximately ninety individuals. The corridor has been designed to include the largest existing fragments between the two larger blocks. It avoids major obstacles — such as the BR 101 highway, a gas pipeline, a railroad, and electric lines — and, wherever possible, follows the forested margins of local streams.

The corridor crosses lands pertaining to private owners, most of whom are engaged in extensive cattle ranching, primarily for meat but also for dairy products. Key threats to establishing and maintaining such a corridor are cattle grazing and fires, both of which tend to eliminate forest species and favor tenacious African grasses, especially the so-called braquiária (*Brachiaria decumbens*) and capim gordura (*Mellinis minutiflora*). Controlling cattle requires expensive fencing. Maintaining fire breaks adjacent to fences is increasingly ineffective in the region, because of the growing frequency and intensity of fires caused by the recurrence of drought in recent years. The latter, in turn, appears to signal a shift in local climatic patterns in response to widespread deforestation — a phenomenon noted in other tropical regions (e.g., Lawton et al. 2001, Laurance and Williamson 2001).

This positive feedback loop caused by widespread forest degradation and deforestation — also observed in frontier areas of the Amazon (Nepstad et al. 1999) — poses an enormous threat not only to forest regeneration but also to the maintenance of existing forest fragments in the region. Furthermore, a forest corridor could act as a conduit for fire, posing a potential threat to forest remnants along its route. Finally, an additional obstacle for establishing and maintaining a forest corridor is that many of the local landowners are "weekend" ranchers from the city of Rio de Janeiro who have adopted extensive grazing as a low-cost and low-maintenance form of land use. Such landowners could be reluctant to assume the high costs and management required by a forest corridor.

In 1996, AMLD established an experimental plantation within the Poço das Antas Reserve, in an area burned in a 1994 fire. On western-facing slopes and similar soils, seedlings of native tree species were planted in two configurations: corridors (30 by 150 m) and circular islands (60 by 60 m). After two years, no differences were noted in the growth or structure of the two configurations, and in both cases nonpioneer species suffered high mortality.

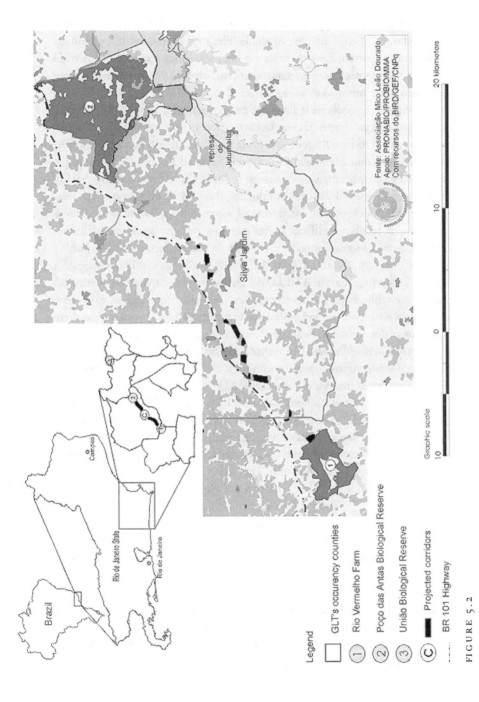

Legend

☐ GLTs occurency counties

① Rio Vermelho Farm

② Poço das Antas Biological Reserve

③ União Biological Reserve

Ⓒ Projected corridors

▬ BR 101 Highway

Fonte: Associação Mico Leão Dourado
Apoio: PRONABIO/PROBIO/MMA
Com recursos do BIRD/GEF/CNPq

Graphic scale

FIGURE 5.2

Forest corridors for lion tamarins. Projected habitat corridor between the Poço das Antas
Biological Reserve and the Rio Vermelho farm in Rio de Janeiro state, Brazil. Source: Golden Lion Tamarin Association.

AMLD concluded that it is more advantageous to plant pioneer species, followed one to two years later by shade-tolerant species after microclimatic conditions are more favorable. Based on these findings, in 1998 AMLD began planting three corridors outside the reserve on properties where tamarins had been reintroduced.

Planting tree seedlings is far more costly (approximately US$3,000/hectare) than natural regeneration, but will increase the probability of quick and successful results, thereby providing a demonstration of corridor implementation for other landowners over an estimated three- to seven-year period. Likewise, rather than attempting to build the entire corridor, AMLD aims to achieve quick success by concentrating initial efforts on a few, relatively short linkages on the properties of supportive landowners. Here landowners will receive seedlings of fast-growing pioneer species and, subsequently, of economic species. To assist in this effort, a major forestry company (Aracruz Florestal) provided 50,000 seedlings of native species[5] for the 2002 planting season. AMLD will also provide the materials needed for corridor establishment — including, most critically, fencing. In exchange, each landowner will provide labor for establishment and subsequent management of the corridor. Additional incentives to attract landowner participation are discussed under the section on economic incentives below.

Determining an appropriate width for the corridor is critical. Brazil's National Environmental Council (CONAMA) defines a forest corridor as a strip of vegetation with a width at least 10 percent of its length. While a convenient guideline, this blanket definition is not appropriate for a wide range of species and diverse environmental conditions. The golden lion tamarin can maintain populations in mature secondary forests — which characterize most of the forest cover in Poço das Antas. Furthermore, evidence from censuses indicates that tamarins are capable of crossing formidable (although short) barriers such as the two-lane BR 101 highway. As a result, a relatively narrow corridor may be sufficient to serve as a conduit for tamarin movement, as long as it provides an axis of intact habitat where edge effects are minimal. Based on these considerations and years of investigations on tamarin behavior, AMLD estimates that a minimum corridor width of 30 meters would suffice for tamarin movement between forest fragments. Long-term maintenance of the corridor in a fire-prone landscape, however, would require a forest strip of at least two to three times that width.

This small-scale habitat corridor for protecting the golden lion tamarin is now part of a much larger-scale conservation initiative. In conjunction with other nonprofit and public institutions, AMLD has begun developing strategies to restore landscape connectivity within the Upper São João river basin and, ultimately, to establish a large-scale landscape corridor connecting Atlantic Forest remnants along a 200-kilometer axis centered in the montane forests of Rio de Janeiro state. Both of these larger-scale efforts would restore key environmental services (such as soil conservation, protection of water quality, and regulation of water flow), thereby benefiting both urban and rural residents.

Stakeholder Engagement

Some of the key stakeholder groups that will determine the success of the corridor are described below.

Conservation Organizations   Conservation of the golden lion tamarin provides an exemplary case of international collaboration. International conservation organizations such as WWF have raised substantial funds and increased public awareness worldwide about the plight of this species. Based on long-term scientific studies, research centers such as the Smithsonian Institution and the University of Maryland have developed simple management techniques designed to increase the likelihood that captive animals released in the wild will survive. Renowned zoos such as the National Zoological Park in Washington, D.C., the Frankfurt Zoological Society, the Copenhagen Zoo, and the Dublin Zoo are also helping to ensure that conservation of tamarins in captivity supports conservation efforts in the wild.

National organizations within Brazil — including its Ministry of Environment, the National Environmental Agency (IBAMA), and WWF Brazil (which is engaged in fundraising efforts and a public awareness campaign) — are assuming increasingly important roles. In addition, Brazilian students have become key participants in research efforts and receive an increasing share of research funding. AMLD has achieved notable success in changing local awareness of and attitudes toward the golden lion tamarin, and it has assumed a coordinating role to ensure that all of these efforts are in tandem.

Today this array of conservation organizations represents a formidable stakeholder group that can provide diverse types of support for establishing a corridor. Such support could include promotion of economic incentives (see section on economic incentives below), technical assistance to landowners, and, potentially, enhanced status for those who participate in this initiative.

Landowners   Through the efforts of AMLD, landowners are increasingly aware of the golden lion tamarin and many are now eager to protect it — and even receive released animals — in forest remnants on their land. This is motivated in part by a growing interest in ecotourism, but the primary motives appear to be increasing appreciation of the tamarin and the status that its conservation confers on landowners (Dietz et al. 1994). These changing attitudes, in turn, reflect intensive environmental education efforts carried out since 1984 by AMLD and, formerly, WWF.

To date, outreach efforts to property owners have yielded extremely positive responses, and five landowners along the planned corridor route are committed to establishing the corridor on their lands. Yet as discussed above there are formidable disincentives for landowners to participate in building a corridor — especially the complex and expensive management measures needed to control cattle (critical while the corridor is being established) and fire (critical during the establishment and long-term maintenance of the corridor). The management issue places AMLD and its allies in a dilemma. While providing materials and technical assistance to establish the corridor is justifiable, landowners must ultimately assume responsibility for management if the corridor is to be sustainable. Engaging landowners in a dialogue to help AMLD understand the specific reasons behind their opposition to the corridor, and developing low-cost measures to address these concerns (such as strategies to diminish fire risk during droughts), should help build support among a core stakeholder group. In addition, AMLD plans to target environmental education to ranch employees because they are in the field on a full-time basis and will be executing corridor management activities.

Formerly Landless Settlers   The federal government settled 104 landless families in a 1,500-hectare area on the southern and eastern edges of the Poço das Antas Reserve in 1994. This and other government-sponsored settlements have helped reduce land conflicts that had festered in the region for years.

Initially this settlement posed a critical threat to the reserve (and its surrounding area, including the proposed corridor route) from wildfires that escaped from shifting cultivation plots and, eventually, hunting and wood extraction. Through an intensive program of community organization and agricultural extension, AMLD is assisting farmer families to minimize the use of fire as a management tool. As an alternative to annual cropping, AMLD has helped establish small nurseries for economic tree species and agroforestry demonstration plots, provided training, and enabled producers to visit other agroforestry projects. In addition, it has carried out soil analyses and assisted in selecting sites and crops for establishing plantations in an effort to promote low-impact land uses. These efforts have led to a significant reduction in the use of fire and expansion of permanent cropping, thereby eliminating much of the threat initially posed by this settlement.

State and Municipal Governments   These stakeholders can develop policy incentives for landowners to take an active role in conserving and even expanding forest fragments in the region (see economic incentives section below) — a critical requirement for the proposed corridor. At present, however, the existing policy framework in Rio de Janeiro provides virtually no incentives for conservation. Changing this framework will require a dialogue with state and municipal governments, as well as other influential decision makers and citizen groups. One such group in which AMLD actively participates is the watershed committee for the São João watershed. Based on a 1997 water resources management act, this and other committees are being set up throughout Brazil to apply increased water tariffs to support water- and watershed-management systems. AMLD has played a central role in laying the groundwork for such a committee for the region in and around the São João river basin, which requires approval of state-level regulations before it can be established formally. In the meantime, an intermunicipal management council has been established that is responsible for developing an environmental management plan for the entire São João river basin, with special emphasis on water. More details on this council are provided in the governance section below.

Land Tenure

Due to past efforts to settle landless people, conflicts over land tenure are now minimal throughout the São João river basin. Nevertheless, technical assistance is needed for newly established settlements of landless people, to ensure that they do not encroach on the Poço das Antas Reserve or undermine corridor efforts. In the absence of effective governmental extension services, AMLD has assumed the primary role in assisting a land-reform settlement near the reserve through the agricultural extension efforts described above. To reduce potential conflicts within this settlement, AMLD also has helped in the delimitation of individual lots and in the definition of equitable water access policies.

Economic Incentives

Economic incentives will be a critical factor in convincing landowners to support corridor efforts. Fundraising is and will continue to be an important source of incentives needed to support corridor establishment costs, technical assistance in corridor management, and purchase of land or easements. The highly charismatic golden lion tamarin provides an effective vehicle for fundraising. Indeed, approximately US$5 million have been raised since 1983 for a wide range of activities related to the conservation of this species. The potential for future fundraising is excellent as a result of the high degree of collaboration among relevant Brazilian and international conservation organizations, as well as increased public awareness of the Atlantic Forest and the multiple threats facing this ecoregion.

Another incentive — which often transcends economic concerns — is the increased awareness of the golden lion tamarin among landowners in the region, largely due to the long-term efforts of AMLD in environmental education. For example, between 1984 and 1992, recognition of the tamarin among sampled respondents increased from 59 to 79 percent (a statistically significant difference), and affirmation of the tamarin's value (either economic or aesthetic) increased from 14 to 62 percent (Dietz et al. 1992). Reflecting these changing attitudes, today landowners are actively engaged in protecting tamarin populations by establishing officially sanctioned, permanent private reserves (abbreviated as RPPN in Portuguese), which increased from zero in 1994 to ten in 2001. Establishing such reserves requires considerable time and expense, in exchange for exempting the area in question from rural land taxes, which are miniscule in Brazil. Some landowners are motivated to establish private reserves because of potential revenues from ecotourism, although currently this is not a major economic activity in the region as the vast majority of tourists ignore inland forests (especially in lowland areas) and flock to the nearby beaches. Many landowners with private reserves have expressed interest in receiving captive-born or translocated tamarins. This interest appears to be motivated by both economic and aesthetic considerations — as well as a new sense of pride in protecting a unique and highly charismatic species.

Another promising development involves watershed committees, such as that for the São João basin and surrounding areas. These could provide a source of revenues from increased water tariffs for water and forest conservation in the region. This development is especially significant because it would be sustainable and not dependent on the good will of donors. However, potential incentives from this source could prioritize hydrologically strategic areas at higher elevations near the watershed source, rather than in downstream areas around the Poço das Antas Reserve.

The state of Rio de Janeiro has yet to adopt the policy incentives for conservation existing in other Brazilian states. For example, the state of Paraná provides an increased stream of tax revenues to municipalities that conserve a higher proportion of their forest cover. Measurement of such cover originally was confined to public protected areas but now includes private reserves. In addition, the process of sanctioning private reserves, formerly under the exclusive control of the federal government, has now been assumed by the state government as well and involves considerably less time and expense. Finally, privately funded efforts are under way in various areas of Brazil to provide revenue streams for carbon sequestration services provided by intact or regenerating forests.

In short, a variety of incentives — both existing and potential — could provide strong motivation for landowners to get involved in corridor efforts.

Governance

The recently established Council for Environmental Management of the São João river basin and surrounding areas represents an important first step in defining and implementing regionwide policies. This council is charged with (i) developing a management plan for the entire São João river basin, with special emphasis on water; (ii) raising funds in support of intermunicipal projects that further the plan's objectives; and (iii) ensuring that municipal projects don't conflict with the plan by keeping open lines of communication with all the municipalities in the basin. Part of the management plan, as stipulated by Brazil's new legislation, should specify appropriate tariffs for water use, which should be reinvested in conservation of the river basin.

The council is composed of representatives from each of the twelve municipalities included in the region, one state representative, and four representatives each from the business and NGO communities (including

AMLD). It will serve as an incubator for future river basin committees for the São João and adjacent basins, which will apply increased water tariffs to support water- and watershed-management systems. These policies will provide critical incentives for conserving and restoring forest cover, thereby reversing land-use trends that have led to widespread forest fragmentation in the region. They could support efforts to establish the habitat corridor planned in the vicinity of the Poço das Antas Reserve and other corridor initiatives contemplated for the region. Defining the roles and responsibilities of the different actors involved in the São João watershed committee is a key issue that is likely to determine the effectiveness of those policies.

Conclusions

This case reveals a number of key issues involving the design and implementation of corridors:

- The 20 kilometer–long corridor now under construction will ultimately enable golden lion tamarins to move between two of the largest forest blocks in the region, following a strategically defined route that makes maximum use of existing forest fragments and water courses.
- This initiative is linked to larger-scale initiatives aimed at establishing forest connectivity in the São João river basin and, ultimately, along a 200 kilometer–long axis in Rio de Janeiro state.
- Efforts to protect the golden lion tamarin have involved an extraordinary alliance of international and Brazilian research institutions, zoos, environmental organizations, and donors.
- Local landowners have become key allies to conservation efforts and will play a central role in corridor establishment and management, yet their support could be dampened by the complexities of managing a forest corridor in a fire-prone landscape.
- The potential threat of an agrarian reform settlement established adjacent to the Poço das Antas Reserve has been minimized through agricultural extension and community organization.
- New water legislation holds the promise of providing substantial economic incentives for forest conservation and restoration, which could support efforts to establish both the habitat corridor under construction and other corridor initiatives.
- A new intermunicipal council for environmental management provides a forum for diverse interest groups, and it could provide a foundation

for supporting and implementing a wide range of efforts to conserve and restore forests in the region.

Case 2: The Talamanca–Caribbean Corridor, Costa Rica

Summary

The Talamanca–Caribbean Corridor encompasses 31,565 hectares of land in southeastern Costa Rica, connecting mountaintops, coral reefs, and numerous ecosystems in between.[6] This landscape corridor is located in one of the most biodiverse areas of Costa Rica and is home to over 90 percent of the country's plant species. While some protected areas lie in the corridor, almost 90 percent of the corridor consists of land that is either privately owned or in indigenous reserves, and much of this area is worked by poor farmers. Land use is mostly unrestricted in Costa Rica, and maintenance of the corridor's forested areas is almost completely dependent on the cooperation of landowners.

Several economic incentives are in place that enable landowners to keep their forests intact. This case study highlights the economic incentives that have been used by the project thus far, and examines the work of a group of local organizations that are working to ensure that these incentives are accessed.

Design and Implementation A site plan prepared by the corridor project stakeholders (defined below) identifies five special elements of biological importance and defines a list of strategies for working with local people to maintain, preserve, and restore habitats within the corridor. The strategies were prioritized based on four criteria: how well the strategy will reduce threats and improve the condition of the target habitat, how far-reaching the strategy is, whether the strategy matches the skills and leadership capacity of the organization, and how much it will cost to implement. Strategies receiving high rankings under all four criteria were given maximum priority for implementation. The site plan's objectives, however, may require further refinement. In addition, the strategies to achieve them are complex and potentially difficult to implement.

Stakeholder Engagement The Talamanca–Caribbean Corridor Commission — comprising local community organizations, indigenous NGOs, and community-development associations — represents the key stakeholders in the

Talamanca region. The commission works with the Costa Rica Department of Forestry to create policies and incentives to benefit the corridor, and it helps local landowners to take advantage of the economic incentives available to them. The local landowners, mostly poor farmers of Afro-Caribbean descent, are the primary stakeholders.

Economic Incentives   The achievements and success of the Talamanca–Caribbean Corridor are dependent on the participation of local landowners. Because most of these are poor and must actively work their land for subsistence, economic incentives are needed to encourage them to keep habitats and forest cover intact. The Corridor Commission is providing the technical assistance that landowners need in taking advantage of incentives for low-impact farming and forest management, carbon offsets, and ecotourism.

Background

The highly biodiverse Talamanca–Caribbean Biological Corridor in southeastern Costa Rica encompasses 31,565 hectares of protected areas, indigenous reserves, and private lands along Costa Rica's Caribbean coast, extending to the Continental Divide at an altitude of 3,820 meters. This landscape corridor includes nine life zones (Holdridge 1978) and encompasses La Amistad and Chirripo National Parks, Atlantic mangrove forests, and Costa Rica's only intact coral reef. It also extends into Panama's 13,226-hectare Bastimentos National Park.

The corridor is one of Costa Rica's biologically richest areas: in it are found 90 percent of Costa Rica's native plant species, more than 350 species of birds, and a variety of large mammals such as jaguars (*Panthera onca*), Baird's tapirs (*Tapirus bairdii*), margays (*Felis wiedii*), ocelots (*Leopardus pardalis*), jaguarundis (*Herpailurus yaguarondi*), and three primate species. The corridor protects aquatic nursery habitat that supports fresh- and saltwater fisheries off Costa Rica's northern Atlantic coast. Its beaches provide nesting habitat for endangered leatherback sea turtles (*Dermochelys coriacea*), and its rivers provide shelter to manatees (*Trichechus manatus*).

While national parks and protected areas secure the corridor's valuable biodiversity and connectivity at high elevations, fragmentation of habitats at the middle and lower elevations is imminent, because of the encroachment of human activities. Most of the residents of the Talamanca region are landless workers or poor landowners, and most people who do own land possess

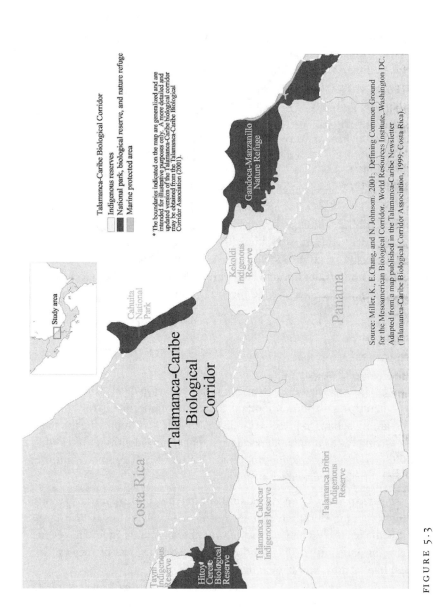

The boundaries indicated on the map are generalized and are intended for illustrative purposes only. A more detailed and updated version of the Talamanca-Caribe biological corridor may be obtained from the Talamanca-Caribe Biological Corridor Association (2001).

Talamanca-Caribe Biological Corridor

Indigenous reserves

National park, biological reserve, and nature refuge

Marine protected area

Source: Miller, K., E.Chang, and N. Johnson. 2001. Defining Common Ground for the Mesoamerican Biological Corridor. World Resources Institute, Washington DC. Adapted from a map published in the Talamanca-Caribe Newsletter (Talamanca-Caribe Biological Corridor Association, 1999; Costa Rica).

FIGURE 5.3

The Talamanca–Caribbean Biological Corridor. Adapted from Miller et al. 2001.

only very small plots and practice swidden agriculture or cattle farming. Many indigenous groups live in the Talamanca region, and Afro-Caribbean people have maintained fishing villages along the Caribbean coast of Costa Rica for approximately two hundred years. More recently, the area has received an influx of poor immigrants from Panama, Nicaragua, and Guyana who have come looking for land and work.

The Talamanca region is one of the poorer, less-developed areas of Costa Rica. Most of its communities lack sewer and waste disposal services, resulting in the pollution of local rivers and streams. Excessive use of fertilizer and pesticides on banana plantations also poses a threat to human health and to the environment. A 1991 earthquake and subsequent flooding exacerbated problems related to poverty and insufficient social infrastructure.

Because the protected areas are interspersed among private lands, it is essential that a conservation plan for the Talamanca region provide economic incentives for local inhabitants to practice conservation. The Talamanca–Caribbean Biological Corridor project (CBTC) — a joint effort of the Costa Rica Ministry of Natural Resources, La Amistad Biosphere Reserve, The Nature Conservancy, and several local NGOs — aims to maintain and expand the corridor through land purchase, agreements with local communities, legal actions, economic incentives, and sustainable development options (CBTC 2001).

Design and Implementation

The CBTC completed a site plan for Talamanca in 1991, with the purpose of identifying the most important conservation priorities in the corridor and designing actions to preserve biodiversity while also improving the standard of living for local inhabitants. The site plan also identifies five natural community types and species groupings of interest in Talamanca and the principal threats associated with each of them:

• *Wet tropical forest.* Unregulated timber extraction and expansion of monoculture cropping (mostly bananas) are major threats to this habitat type. These activities also precipitate a variety of other problems, including severe soil erosion and nutrient loss, disturbance of animals, loss of species, decreased habitat area, and habitat fragmentation.
• *Coastal wetlands.* These areas are also endangered by timber harvest, and by drainage for conversion to aquaculture. This habitat type includes "yolillo" palm (*Raphia taedigera*) forests, which are now restricted to just a

few locations, as well as the only protected mangroves in Costa Rica. Biodiversity loss and sedimentation are severe in Talamanca's coastal wetlands.

• *Rivers, lagoons, and ravines.* Freshwater habitats in Talamanca face problems such as overfishing, diversion of water for agriculture, pollution from agrochemical runoff and other sources, and alteration through loss of forest cover and sedimentation.

• *Large mammals.* Species such as tapirs, jaguars, pumas (*Puma concolor*), sloths (*Choloepus* spp.), and anteaters (*Myrmecophaga tridactyla*) are threatened by habitat fragmentation and also by hunting. Hunting of small species such as the agouti (*Agouti paca*) and collared peccary (*Tayassu tajacu*) for sale in the pet trade is also problematic because it removes food sources for some large mammals. The CBTC is interested in connecting nature reserves in the Talamanca region in order to create habitat blocks large enough to support viable populations of large mammals. It also is interested in using large mammals as umbrella species for corridor design because viable populations require large areas of good-quality habitat.

• *Migratory birds and forest-dependent birds.* Bird species are also severely impacted by loss of forest habitat and fragmentation. The CBTC recommends monitoring movements of altitudinally migrating birds as one way of identifying corridors of intact habitat linking highland and coastal areas.

Most landscape corridor initiatives define a broad set of objectives. But the above list may require further refinement before it can serve as a strategic set of achievable targets.

The CBTC has defined six objectives as highest priority for implementation.

• *Promote environmental education.* Education in ecology, sustainable natural resource use, and the importance of habitat conservation and biodiversity is needed at all levels in the communities of the Talamanca region. This involves hiring and training teachers from local communities. CBTC has been doing environmental education work in fourteen schools in Talamanca since 1999.

• *Lobby for changes in environmental policy.* One of the greatest weaknesses of environmental protection policy in Costa Rica is the fact that many existing laws do not apply to current environmental problems, or are not flexible enough to deal with evolving threats. The regulations needed to address currently urgent issues — such as protection of endangered and endemic species and fragile ecosystems, or regulation of extractive activities — either do not exist or are not effective. To promote better environmental policy, CBTC will rely on field staff with broad experience in environmental law and good working relationships with local environmental groups. Five-year policy goals include better protection of rare and endemic species, management zoning for wetlands and mangrove forests, and better support for personnel needed to enforce environmental laws.

• *Change forestry policy.* A number of problems with forestry policy in the Talamanca region need to be addressed. Many forest species are disappearing, and no technical criteria currently exist for forest management. While timber extraction is one of the only viable employment options in rural areas, the harvest-permitting process is not rigorous or consistent, and local communities receive only a small share of the price for processed timber. Changes in forestry policy will be necessary if existing forests in the region are to be managed for long-term conservation. CBTC is lobbying for improvements in forestry policy, using current information on Costa Rica's forestry sector. Some specific policy actions being encouraged include:

• support of better research on forest ecosystems;
• legislation to protect threatened species;
• education of forest owners on the nonextractive value of forest resources;
• establishment of an effective registry for tracking the use of forest lands; and
• creation of a state and local system to monitor the illegal cutting, sale, and shipment of timber.

• *Help landowners receive payments for ecosystem services.* The Costa Rican government has a program that compensates landowners for the ecosystem services provided by intact habitat, including reduced emissions through carbon sequestration. Unfortunately, the certification and registration processes that are required to qualify for benefits are complex and expensive. CBTC is providing funds and technical assistance to help landowners receive benefits. CBTC places this among the highest-priority strategies, as it will ensure that sustainable land use will continue in the future. Helping Talamanca landowners obtain payment for ecosystem services also brings added benefits, including opportunities for CBTC to meet and share information with local stakeholders, and thereby gain exposure to different properties throughout the region. Such exposure will help identify conservation priorities at a regional scale.

• *Research.* More field research is needed to catalog the animal and plant diversity and to define the habitat requirements of organisms in the Talamanca region. To promote research efforts in the area, CBTC plans to establish a consortium of Costa Rican government agencies and national and international universities.

• *Develop a fundraising program.* Fund raising is essential for long-term implementation of all conservation strategies. CBTC has been raising funds since its inception in 1992 by applying for grants from diverse institutions. Funding provided by the GEF-UNDP expired in 2001, creating an urgent need for new sources of support.

Six additional intermediate-priority objectives have been defined by the CBTC.

• *Use local people for monitoring and enforcement.* Hiring local people to monitor and enforce environmental regulations can provide a source of income to local communities and help deepen identification with and pride in local resources. The key to the success of this strategy is in knowing the communities and identifying effective community leaders. Ideally, guards will help protect wildlife and collect data to increase the knowledge base for scientific management.

• *Develop a database to monitor logging permits.* The Costa Rican Department of Forestry (MINAE) needs assistance in developing an up-to-date database that will aid the effective monitoring of logging permits. Existing permits and other restrictions are currently very poorly enforced, in part because resource managers lack the information needed to effectively track timber harvesting, processing, and sale. CBTC is creating GIS maps to be used for this purpose.

• *Strengthen ecotourism.* Ecotourism has been increasing in the Talamanca region, and the industry could cause damage it if not properly regulated. Potential problems created by ecotourism include generation of excess garbage and poorly planned facilities construction that damages wetlands and reefs along the coast. Tourism in the Talamanca region to date has focused disproportionately on beach-based activities, limiting its potential to encourage protection of inland forest resources. Asociación ANAI, a Costa Rican NGO that participates in and is supported by the CBTC, is developing a plan to promote ecotourism at a scale appropriate to local conditions. Tourism development should be designed to provide income for local communities while protecting natural resources. As a start, CBTC plans to support initiatives that enable local people to develop tourism programs, build infrastructure, and dispose of or recycle garbage.

• *Diminish the environmental impacts of agriculture.* Most local farmers currently practice swidden agriculture and subsidize their income from livestock and crops with timber sales. CBTC suggests that farmers could improve their incomes and use resources in a more sustainable manner by switching to higher-value organic crops or hydroponics. Farmers could also use polycropping systems like those employed by many indigenous people, in which multiple crops are planted together, generating less stress on both crops and soil by mimicking the structure and function of natural biotic communities. In addition, CBTC wants to encourage farmers to reduce their use of agrochemicals and increase tree planting to reduce agriculture impacts on surrounding forests. CBTC is collaborating with the Talamanca Agency of Small Producers (APPTA) to promote sustainable agricultural practices.

• *Increase payments for environmental services.* While an existing program already allows landowners to receive payments for the environmental services provided by some types of undeveloped land, it is extremely limited in scope. This program should be expanded to include payments for mature forests, areas of aquifer recharge (key to watershed protection), and sites that are critical for connectivity. CBTC has already identified many of these key sites through a rapid ecological assessment, and hopes to coordinate efforts with Costa Rican forestry officials to promote their protection through economic incentives.

• *Monitor impacts of banana plantations.* Banana plantations have destroyed extensive wildlife habitat in the Talamanca region and also produce large amounts of garbage and chemical pollution. A combination of positive incentives for companies owning plantations that comply with standards such as those defined by ISO (International Organization of Standardization), together with application of negative incentives such as fines for companies that do not, could ameliorate this situation.

Finally, the CBTC has defined four lower-priority objectives:

• *Identify and map mangrove habitats so that they can be protected.* Costa Rican environmental law already provides for the protection of mangroves. Yet in the Talamanca region most mangroves have not been mapped, and protection of these habitats is inadequate. As a result extensive areas of this ecosystem have been dredged and drained for agricultural use.

• *Restore degraded forests.* Restoration is needed for many of the forested lands in the Talamanca region. This is particularly important along the coast and in riparian areas that are important for the control of flooding and protection of aquifer recharge zones. CBTC plans to build a nursery to supply seedling trees.

- *Create wildlife farms for species threatened by hunting.* Although this strategy could be of importance to the survival of some species, it is not a high-priority action because CBTC lacks experience in this area. In addition, startup costs are likely to be high, and its success in protecting wildlife is uncertain.
- *Enforce existing standards for road construction.* Standards for road construction in forests must be enforced to prevent the excessive erosion that is caused by poorly constructed, substandard roads.

The extensive nature of the above list suggests that the site plan devised for the Talamanca–Caribbean Corridor may have an excessive number of strategies, thereby increasing the complexity of its implementation. Furthermore, many strategies appear to have dubious value in relation to their capacity to reduce threats and improve the condition of the target system, and to address multiple problems. For example, the expansion of agriculture and its negative environmental impacts represent a formidable threat to the corridor, yet the CBTC considers that addressing this threat is an intermediate priority, on the same level as tourism and its potentially negative impacts (which currently are limited to beach areas). In the meantime, environmental education and fundraising are considered to be top priorities — probably reflecting the skills of the CBTC's members but not necessarily the most urgent issues facing the corridor.

Stakeholder Engagement

Almost 90 percent of the Talamanca–Caribbean Corridor is made up of private land and indigenous reserves, with the small remainder protected in national parks (Rafael Calderón, personal communication). Land use on privately owned plots is only partially restricted in Costa Rica, so progress in establishing connectivity in the corridor has been slow and somewhat difficult.

Work by a small group of Costa Rican NGOs on the development of the Talamanca–Caribbean Corridor began in 1990, and today between fifteen and nineteen organizations are participating members of the CBTC. Members include local NGOs such as Asociación ANAI, community-based nonprofit groups made up of farmers and indigenous peoples, and community-development associations. The Costa Rican Department of Forestry is an unofficial member of the commission, advising policy decisions without having a vote or a place on the board of directors. The commission remains, therefore, a private or community-based organization that must rely on a consensus process to keep stakeholders on board, with no enforcement power to put behind its management recommendations.

The success of the Talamanca–Caribbean Corridor project is highly dependent on the participation of local landowners, and the CBTC works with them to find economic incentives for maintaining forest cover and managing land in a sustainable way. The CBTC makes a special effort to identify and work with landowners whose property includes habitat of particular biological importance. Yet it also seeks the involvement of as many landowners as possible, regardless of the quality of their holdings.

One of the major organizations working in the region is The Nature Conservancy, which considers the Talamanca–Caribbean region to be a top priority for conservation and has been supporting local NGOs and other participants through its Costa Rica field office. TNC provides training, technical assistance, and research support aimed at corridor development. It has provided rapid ecological assessments for marine and terrestrial resources, and helped CBTC to develop the site conservation plan. TNC funding for the Talamanca work comes from the Adopt an Acre Program, the Parks in Peril Program, and from private funding.[7]

Economic Incentives

Local landowners in the Talamanca region are the ultimate decision makers in the corridor project, and their willingness to maintain forest cover and manage their land soundly will determine the fate of most of the existing natural habitat in the region. Several economic incentives exist for landowners to manage their land as forest. CBTC and its member organizations are working with local people to identify these incentives and benefit from them.

• *Organic agroecosystems.* In moist areas of Costa Rica, high-intensity use of small plots of land can generate a higher rate of return than when the land is managed for cattle production (Diego Lynch, personal communication). CBTC provides technical assistance to local landowners to develop high-intensity schemes in the form of organic cocoa plantations. Because cocoa grows in shade, plantations in forest mimic natural ecosystems and harbor forest-dwelling wildlife.[8] By helping local farmers to manage their land for cocoa production rather than cattle ranching, landowners receive a greater return on their investments and, at the same time, connectivity in the corridor is maintained or restored.

- *Sustainable forest management.* Landowners can benefit from leaving their forests standing by extracting forest products in a sustainable manner. In the Talamanca region many farms and indigenous lands are managed as forests, from which landowners or community members extract wood, medicinal plants, and crafting materials for sale in local markets. CBTC provides technical assistance to landowners wishing to manage their land in this way, demonstrating sustainable forest management and extraction practices.

- *Ecotourism.* While this is still in very early stages, some communities in the Talamanca region have begun using ecotourism as a source of income from otherwise undeveloped forest areas. CBTC is working with community groups to set up lodges and guide associations, and is helping them to set aside earnings from ecotourism to benefit local conservation initiatives. Lodges have begun to receive their first guests only recently, and it is too early to judge the success of ecotourism in the area. Nevertheless, the success of ecotourism as a viable alternative to deforestation would create strong economic incentives for landowners to keep their forests standing (Diego Lynch, personal communication).

- *Compensation for environmental services (carbon offsets).* Landowners who participate in conservation activities such as reforestation, natural forest management, or forest preservation can receive payments from the Costa Rican National Forestry Fund (FONAFIO) of US$42–105 per year per hectare of standing forest (Malavasi and Kellenberg 2002). Payments are funded primarily through a 5 percent sales tax on fossil fuels in Costa Rica, which generates enough money to compensate landowners of approximately 3,000 hectares of forest throughout the country. Although limited at present, this program eventually could provide an important incentive for poor landowners to keep their forests intact. Unfortunately, few poor people have been able to take advantage of the program. The complex and costly procedures required to obtain these incentives are prohibitive for the poor, and only those wealthy landowners with political or government connections have been able to access the available funds. Some of CBTC's member organizations work with poor people to help them gain access to FONAFIO.

- *Land purchase.* While most landowners choose to farm their lands, some find that they can sell it to conservation organizations for more than they can make on it in a year. Some privately owned land in the Talamanca–Caribbean Corridor has been purchased by conservation organizations in this manner, and these plots are likely to be preserved as standing forests for the long term.

Conclusions

An impressive array of local, national, and international organizations has joined forces to design and implement the Talamanca–Caribbean landscape corridor. The objectives of the corridor site plan, however, are unclear, the strategies complex, and the priorities questionable. This may reflect, in part, the diversity of stakeholder interests involved. Because most of the region consists of private lands or indigenous reserves, the plan calls for a focus on economic incentives, which will be critical for the long-term success of the corridor.

Case 3: A Corridor Network for Wildlife in Florida, USA

Summary

This case focuses primarily on the incipient Florida Panther Corridor, which is designed to facilitate expansion of highly endangered Florida panthers (*Felis concolor coryi*) from southwestern Florida to available, privately owned lands to the north.[9] The Florida panther's population has remained under 100 individuals for several decades, is highly inbred, and has limited dispersal potential due to landscape barriers (Maehr et al. 2002). Once complete, the corridor would protect some of the best remaining privately owned habitat for Florida panthers, preserve a mosaic of unique habitats supporting a variety of other species, and also protect important watersheds. This case illustrates how a conservation organization can bring together key stakeholders (i.e., landowners and state natural resource agencies) to secure conservation land. The Florida Panther Corridor also provides an example of ongoing corridor implementation through the pursuit of local conservation opportunities within the context of a landscape-scale corridor vision. Conservation easements and various economic incentives are additional techniques used to secure conservation on private lands. Finally, the case provides insight into how governance and land tenure issues are being resolved in practice.

Background

Since 1994, TNC has spearheaded efforts to implement biological corridors that have been planned by several state governmental entities in the

U.S. state of Florida (e.g., Cox et al. 1994, Florida Department of Environmental Protection and Florida Greenways Coordinating Council 1998, Hoctor et al. 2000), as well as by their own organization (Core Technical and Planning Team 2001a and 2001b). The long-term goal is to establish a landscape corridor of predominantly intact wildlife habitat extending from Everglades National Park northward to conservation areas in southeastern Georgia (i.e., Okefenokee National Wildlife Reserve) (fig. 5.4).

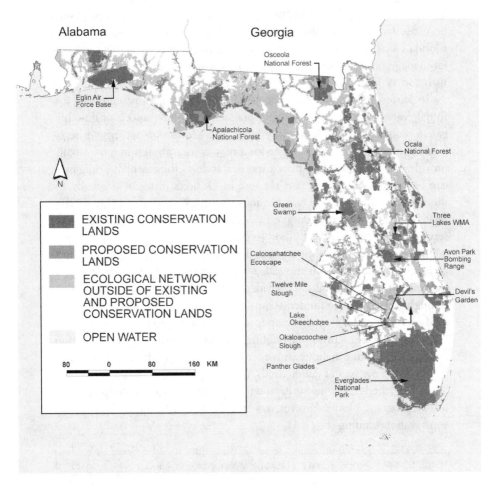

FIGURE 5.4

A landscape corridor for Florida. Map shows new corridor projects that propose to link current populations of Florida panther on public lands in southern Florida to suitable habitats in central Florida. Adapted from Hoctor et al. 2000.

Working with numerous state and federal agencies, and utilizing several existing conservation plans, TNC identifies strategically located, high-quality habitat, often through aerial photographs and Landsat satellite imagery. TNC then contacts landowners to gauge their interest in entering into one of several kinds of conservation agreements. With the landowner's permission, it conducts biological assessments of all natural community types on the property, utilizing field and aerial surveys, ground truthing, and recording of GPS coordinates for rare species.

If owners are willing to sell or place conservation easements over their land (boxes 5.1 and 5.3), TNC then prepares a formal application proposing purchase by state (e.g., the Florida Forever program administered by the Florida Department of Environmental Protection, one of Florida's five water-management districts) and, occasionally, federal agencies (e.g., the U.S. Fish and Wildlife Service).

To hasten the acquisition process, the Conservancy may also pay the initial costs for boundary surveys, appraisals, and other aspects of due diligence required for purchasing the land. The acquisition agreement negotiated between TNC and the state for a new conservation purchase usually provides reimbursement for these expenses. Indeed, the cost of due diligence can be quite expensive, in part because of Florida's stringent standards. In the case of the 23,221-hectare Pinhook Swamp Corridor and the 50,610-hectare conservation network it created, for example, the total cost of these services approached US$1 million (see box 5.4).

In the recent past, public funding for acquiring new conservation lands in Florida has come from the state's Preservation 2000 (P2000) program. Established in 1991, P2000 provided US$300 million per year, which through early 2001 had funded acquisition of over 500,000 hectares.[10] TNC has helped the state invest this money in conservation projects by identifying, proposing, and sometimes negotiating the acquisition of key conservation lands. TNC-led negotiations and other initiatives helped to encumber up to $100 million per year of P2000 funds, typically focusing on large, landscape-level, watershed-based, and critical wildlife corridor projects. Since 2000, a similar program, Florida Forever, has provided the vast majority of the state's conservation funding (box 5.1).

BOX 5.1. THE FLORIDA FOREVER PROGRAM

In 2000, the Florida state legislature established a new program, Florida Forever, to replace P2000. Like P2000, Florida Forever is designed to provide

continued

US$300 million/year for ten years, but Florida Forever gives increased emphasis to projects that include water-resource supply (including aquifer recharge zones), watershed protection, strategic conservation linkages, and restoration of ecological systems.

Florida Forever introduced new documentation requirements to increase accountability. It also established a new advisory board (the Florida Forever Advisory Council) charged with overall program oversight, direction, and legislative reporting, and an expanded Acquisition and Restoration Council (ARC) that approves and ranks many of the projects for funding. Besides introducing additional proposal requirements, the new program has a large number of acquisition criteria, goals, and measures that must be addressed. Proposals that meet the greatest number of these will, theoretically, rank higher and receive funding sooner. In spite of the increased requirements and broader programmatic scope associated with these changes, Florida Forever appears to provide an effective approach to protecting landscape-scale areas required for conserving biodiversity, maintaining ecological integrity, and sustaining functional ecosystems.

Design

The purpose of the Florida Panther Corridor is to link federally protected lands supporting Florida panthers in protected areas of southwestern Florida (Big Cypress National Preserve, Everglades National Park, and the Panther National Wildlife Refuge) to prime panther habitat located on private lands in central Florida. Parts of the corridor route trace the path used by three radio-collared Florida panthers that crossed the Caloosahatchee River into south-central Florida during the past few years. The main objective of the corridor is to facilitate the dispersal and establishment of a new, viable, and breeding subpopulation of Florida panthers in central Florida (fig. 5.4).

One of the primary goals of the Florida Panther Corridor is to link federal conservation lands in southwestern Florida to the 64,516-hectare Fisheating Creek Ecosystem project, also proposed for protection by TNC to the state of Florida. With the subsequent connection of the Fisheating Creek project to other areas (including the Lake Wales ridge, an important area for many federally listed plant species, Lake Okeechobee, and the Kissimmee river valley), a large conservation landscape can be established that should facilitate movement of Florida panthers from southwestern Florida to suitable high-quality habitat in central Florida.

Human activities have generated substantial impacts in parts of the Florida Panther Corridor. Some tracts have been logged but allowed to regenerate as natural pine stands. In other, smaller areas, intensive agriculture — primarily involving citrus, sugarcane, and other row crops — has caused localized habitat conversion and alteration of regional hydrology. Every effort, however, has been made to exclude these kinds of land uses from the corridor. Commercial cattle ranches may also impact native habitat through conversion and simplification of natural communities. Located within 65 kilometers of such rapidly growing cities as Fort Myers, Bonita Springs, and Naples, the corridor is also under increasing pressure from residential and commercial development.

Corridor habitats include a mosaic of several natural community types embedded in a matrix of mesic and wet flatwoods that are characterized by a South Florida slash pine overstory and a dense, shrub-dominated understory. The other predominant habitats include cypress-dominated strand and dome swamps; live oak– and cabbage palm–dominated prairie hammocks; grass- and sedge-rich marl prairies; diverse, herb-dominated wet prairies; and occasional depression marsh systems scattered throughout the flatwoods.

While preserving the highest-quality habitat available is usually the first priority when acquiring conservation lands, other factors — such as actual use by target species — also become key when a corridor's primary objective is to provide linkage for species movement. The Florida Panther Corridor includes some segments of less than ideal habitat, such as improved pasture and occasional agricultural fields, and it spans several paved highways. In some cases, this lower-quality habitat provides the only available linkage for Florida panthers, and radio tracking has shown that panthers will cross such areas. As a result, connecting a series of more natural habitats through occasional disturbed parcels, while a concern, is not seen as a major obstacle to corridor success — especially given that habitat restoration for some of these areas is also planned.

The Florida Panther Corridor habitats also support other species of conservation interest, including American alligator (*Alligator mississippiensis*) and Florida black bear (*Ursus americanus floridanus*). Several threatened or endangered birds, including Wood Stork (*Mycteria americana*), Crested Caracara (*Caracara cheriway*), Florida Sandhill Crane (*Grus canadensis pratensis*), Florida Scrub-Jay (*Aphelocoma coerulescens*), and southern Bald Eagle (*Haliaeetus leucocephalus*) also inhabit the proposed corridor area. Conservancy ecologists also documented two rare plant species — hand fern (*Ophioglossum palmatum*) and mock vervian (*Glandularia maritima*) — within the proposed linkage areas (Hilsenbeck and Caster 1999a, 1999b).

In addition to supporting wildlife, the Florida Panther Corridor assists in the protection — and potentially the ultimate restoration — of the watersheds and complex hydrological regimes of southwestern Florida. Water flowing through the sloughs of the proposed corridor feeds the Fakahatchee Strand subtropical forest area, Big Cypress Swamp, and other portions of the western Everglades ecosystem. The Florida Panther Corridor also provides essential protection for the headwaters of several watersheds, through which fresh-water flows to coastal estuarine nurseries that support rich commercial and sport fisheries in the Ten Thousand Islands region (Hilsenbeck and Caster 1999a, 1999b). In short, the corridor's purpose is far greater than just to protect panthers and other species, for ultimately it will conserve ecological integrity and ecosystem function over an immense area.

The costs of securing private lands for corridors or other conservation projects, whether through fee-simple purchase or conservation easement, vary greatly depending on the location of the property. Properties near urban areas have greater development potential and tend to be most expensive. Highly rural areas and areas that are too wet to convert easily to farming or other development typically cost less. Main et al. (1999) calculated the cost for purchase and management of four existing conservation areas in southern Florida at US$69/hectare/year, paid in perpetuity. This cost is based on an average purchase price of $1,291/hectare (range US$788–6,250/hectare), assumes an interest rate of 3.65 percent (the thirty-year U.S. treasury bond rate in constant US$), and does not include transaction costs. However, privately owned priority panther habitat just to the north generally commands a higher fair-market value, due to its higher potential for agricultural and residential development. The mean value for 187,201 hectare of agri-culturally suitable land in this area, excluding parcels near urban bound-aries, was found to be $4,744/hectare (Main et al. 1999). This value appears to be representative of land costs in the Florida Panther Corridor, although as noted above these costs vary considerably. Using this value, Main et al. (1999) calculated the cost for purchase and management of privately owned lands in southwest Florida at US$200/hectare/year, paid in perpetuity.

Significant progress has been made in the last seven years in the design, implementation, and initial acquisition of corridor lands through the devel-opment of various projects. Several factors make it necessary to build this landscape-scale corridor in a piece-by-piece fashion. While TNC identifies and prioritizes properties according to habitat values and the role they can serve in the landscape-scale corridor, the order in which tracts are acquired and the final configuration of the corridor will depend on the availability of willing sellers, the ability to negotiate a fair-market price for the lands, and funding opportunities. The process of contacting landowners, gaining their trust, completing ecological surveys, and proposing lands for conservation is time-consuming. In addition, enough owners with the right mix of contiguous lands must be brought together to form a coherent project that can be proposed for acquisition by the state. As a result, the Florida Panther Corridor is being built deliberately, but on a project-by-project basis. Major projects include the following (see locations in fig. 5.4).

Okaloacoochee Slough   This 11,893-hectare area protects a substantial portion of the watershed and headwaters that feed the Fakahatchee Strand subtropical forest, provides a movement corridor and high-quality habitats for Florida panthers and Florida black bears, and includes a portion of the route used by radio-collared panthers to travel north to the Caloosahatchee River. TNC worked with the landowner (a major agricultural interest), submitted the original proposal, and helped to position the project for purchase by the South Florida Water Management District in 1995. The Florida Division of Forestry now manages this property.

Caloosahatchee Ecoscape   According to the U.S. Fish and Wildlife Service and the Florida Fish and Wildlife Conservation Commission, this 7,173-hectare area provides *the* critical linkage for Florida panther movement across the Caloosahatchee River. Three radio-collared Florida panthers utilized the habitat within the area to disperse northward into the adjacent Fisheating Creek Ecosystem project. In addition, at its southern end the Caloosahatchee Ecoscape connects to the northern portion of the Twelve Mile Slough project (see below).

At the time the Caloosahatchee Ecoscape was first proposed for protection, the state council overseeing the largest portion of the P2000 funds scaled back the original 12,000 hectare proposal, believing there was insufficient evidence that Florida panthers utilized the area. Within two years, data were obtained showing that panthers crossed the Caloosahatchee River by moving through the exact lands encompassed by the project, but by then the excluded eastern portion had already been converted almost entirely to intensive, row-crop agriculture. Even with new data showing the area's importance, no tracts have been acquired to date. This may be due to the fact that, despite its significance as the last currently viable linkage for panthers moving northward in south-central Florida, the Caloosahatchee Ecoscape also contains less pristine habitat than is usually prioritized for state funding.

Panther Glades   This 7,967-hectare tract — encompassing parts c the Kissimmee Billy and Tony strands and Lard Can Slough of southwest Hendry (and adjacent Collier) County — includes some of the best remaining core habitat for Florida panthers. The property is divided among eleven private landowners, families, or corporations. As of October 2001, much of the land was under appraisal and negotiations were under way for both the fee-simple and less-than-fee (i.e., conservation easement) acquisition of the five largest properties. In May 2002, the opportunity arose to purchase an additional 8,459-hectare property, the Dinner Island tract, and the Panther Glades proposal was amended to include this piece, which was available for fee-simple purchase. As of June 2002, the Dinner Island tract had been acquired, partially because its acquisition was not slowed by the often-protracted negotiations that surround conservation easements. It is now managed by the Florida Fish and Wildlife Conservation Commission.

Twelve Mile Slough   This 5,310-hectare tract contains a mosaic of high-quality habitats similar to the Okaloacoochee Slough. The core of the area is a major tributary to Okaloacoochee Slough, and it therefore also contributes substantial flows to Fakahatchee Stand and the Ten Thousand Islands estuarine system. The several ownerships in the project are under appraisal for fee-simple purchase (with one 3,035-hectare tract already acquired) and, if the Conservancy is successful in obligating state funds, these acquisitions will connect Okaloacoochee Slough State Forest to the currently unprotected Caloosahatchee Ecoscape area.

Stakeholder Engagement

Stakeholders in the Florida Panther Corridor include numerous private landowners and at least four major corporations. Agencies involved at the state and local level include: (i) the Florida Forever/Board of Trustees program (formerly known as the Conservation and Recreation Lands [CARL] program) within the Florida Department of Environmental Protection; (ii) the South Florida Water Management District; (iii) the Hendry, Collier, and Glades County Commissions; (iv) the Florida Division of Forestry; and (v) the Florida Fish and Wildlife Conservation Commission. Federal natural resource management agencies, such as the U.S. Fish and Wildlife Service, are also involved.

Most of the land in the Florida Panther Corridor is divided among twenty major landholders, primarily ranchers and citrus and sugarcane growers or the corporations that oversee these activities. TNC directly approaches these landowners to ask if they would be interested in selling land or perpetual conservation easements to the state. A number of landowners have been willing to sell their lands when they retire or downsize, or to sell conservation easements because they have a genuine interest in conservation. Other land-owners and, initially, the Hendry County Commission, have been less en-thusiastic about considering these options. Corporate landowners particu-larly are often reluctant to participate in conservation agreements, although they recently have relaxed their positions and allowed TNC to propose some of their lands to the state. Indeed, two corporate owners have already sold key tracts in the proposed corridor to the state.

A land swap may be arranged to attract one remaining landowner, a corporation that owns the majority of a critical tract needed to implement the entire corridor. This tract would close a 2.4-kilometer gap along the corridor route, fronting a major highway on which several Florida panthers have been killed over the past ten years. The Florida Department of Trans-portation will not consider building a wildlife underpass (box 5.2) until property on both sides of the road is secured under a conservation agree-ment. To bring this landowner into the process, TNC and the state of Florida hope to arrange a land exchange that would trade 200–800 hectares of orange groves and sugarcane on Dinner Island for the relatively intact hab-itats owned by the corporation.

BOX 5.2. UNDERPASSES FOR FLORIDA PANTHERS

Collision with motor vehicles is a major cause of wildlife mortality worldwide. Many techniques have been tried to decrease roadkills, but the most successful and widespread involve underpasses combined with roadside fencing. The most ambitious of these projects to date is a series of twenty-four underpasses along a 64-kilometer fenced portion of the four-lane I-75 highway traversing the Florida Panther National Wildlife Refuge and other conservation areas in southwestern Florida. Each underpass costs US$350,000 and consists of two crossings, 37 me-ters long and 13 meters wide, divided by an open median that is 22 meters across. Remote cameras sensitive to movement have documented use of the underpasses by Florida panthers, Florida black bears, bobcats (*Lynx rufus*), white-tailed deer, otters (*Lutra canadensis*), raccoons (*Procyon lotor*), and turkeys (*Meleagris galla-pavo*), and frequency of use for many species (including Florida panther) is in-creasing over time. A simpler and less costly (US$128,000) underpass — consisting

continued

of a precast, concrete box culvert 2.4 meters high, 7.3 meters wide, and 14.6 meters long—has been installed along a smaller, two-lane highway (SR 29) and has been found to be used by the above species and, in addition, by gray fox (*Urocyon cinereoargenteus*) and five species of wading birds.

—Sources: Foster and Humphrey 1995, Lotz et al. 1997, Havlick 2003

Even when landowners are willing to sell or otherwise conserve their properties, acquisition of land or conservation easements involves a lengthy consensus-building process among the various stakeholder groups. For example, in some instances finalizing the myriad terms of a perpetual conservation easement has taken up to two years of careful and painstaking negotiation. This process entails not only establishing the terms of the agreement but appraising and then, in essence, re-appraising the land to determine by how much specific land-use restrictions will diminish its market value.

The Hendry County Commission initially opposed state purchase of the Caloosahatchee Ecoscape and required additional incentives to endorse the project. Hendry is one of the most rural counties in southern Florida and has a predominantly agriculturally based economy. As a result, local officials understandably tend to prioritize development and economic growth over land and wildlife conservation. Conversion of private land to public lands, such as state forests, reduces the local tax base and can deprive the county of needed revenue. However, the Florida Division of Forestry has a long-standing policy that 15 percent of all timber revenues from a state forest are provided to the county. In addition, P2000 and Florida Forever have provisions providing payment in lieu of taxes for lands acquired by the state to help defray losses of property tax revenues to local governments and school boards. Opportunities for recreation development through Florida's Rails-to-Trails program provided an additional incentive. (See section on economic incentives below.) Once Hendry County officials became aware of these programs, they became more willing to negotiate.

State and federal natural resource agencies generally have been highly supportive of initiatives to acquire land for the Florida Panther Corridor project. Thorough and well-documented proposals and site visits to the proposed conservation areas have helped TNC and other environmental NGOs (such as the Florida Wildlife Federation) to convince agency officials and their representatives that the Florida Panther Corridor provides a strategic approach to conservation in the region.

Land Tenure

TNC is working to build the Florida Panther Corridor through a combination of fee-simple and conservation easement purchases by the state.
More than half the land proposed so far (65,685 hectares) will be purchased
outright from people who are interested in retiring or downsizing their holdings, or who feel that portions of their lands are too wet to farm or ranch
effectively. Conservation of the remaining lands is being sought through
purchase of perpetual conservation easements from owners who want to
continue owning and using their properties.

Land tenure shapes how land is used and what natural resource benefits
may be derived from it. Virtually all of the land negotiated in fee-simple by
the Conservancy on behalf of the state of Florida will remain open to the
public in the form of state forests or wildlife management areas. Low-impact
recreational activities such as hiking and canoeing will be allowed on many
public tracts, and hunting on many others. Based on an assessment of the
property and the previous owner's management philosophy, TNC provides
recommendations to the various funding programs and agencies about recreation activities that are compatible with conservation.

Conservation easements provide an alternative to fee-simple purchases.
Under a conservation easement, the purchaser acquires the property's development rights, usually to ensure that current land uses will not change.
Depending on the easements' specific terms and conditions, in most cases
owners continue to use their property and maintain existing structures and
land uses, but they may not expand these activities to convert more natural
habitat or initiate new improvements (box 5.3). At the same time, however,
TNC and the state of Florida recognize that owners need to retain some
flexibility in their activities in order to make a living from and remain on
the land. Most private lands under easement are not open for public use,
but the easement agreement usually specifies limited access for scientific
research and public education.

BOX 5.3. CONSERVATION EASEMENTS

In cases in which landowners wish to continue owning and using their land,
TNC may help negotiate a conservation easement to allow a state or federal
agency to purchase development rights associated with the land (less than fee-
simple). In addition to securing the conservation benefits of land for less than

continued

the full fee-simple appraised value, easements allow the state to save on management costs because the landowner continues to manage the land.

The cost of conservation easements varies according to the location of the land, its current and potential uses, and the degree and kind of land-use restrictions the easement entails. Using a mean land value of $4,744/hectare, Main et al. (1999) calculated the cost for purchase and management of easements in priority panther habitat in southwestern Florida at US$116/hectare/year, paid in perpetuity — just over half the cost required for fee-simple purchase (US$200/ hectare/year). State-approved appraisers assess the value of land according to its "highest and best use," which is a traditional valuation term in the industry. For example, a conservation easement that restricts future development options must compensate the owner for the loss of value associated with those development rights. Lands with higher development potential are more expensive, and an easement over these lands demands a higher percentage of the total appraised value to purchase development rights. Conservation easements for rural properties in Florida with less immediate development potential have been purchased for as little as 28 percent of appraised value, while one easement in an urban area with high development potential cost 98 percent of appraised value.

By allowing the purchase of development and other rights, conservation easements typically ensure that additional habitat will not be converted to other, more intensive uses. Landowners in Florida can usually maintain existing houses, barns, fences, roads, wells, and other such improvements. Many continue to use their land for agriculture, including livestock grazing, limited timber harvest, and hunting of game species. Existing land-management strategies, including prescribed fire, are also explicitly encouraged. Easements, however, typically prohibit the expansion of agricultural intensification, mining, further development, and all other such uses that are not compatible with conservation purposes.

In Florida, landowners who wish to sell or grant (i.e., donate) a conservation easement must also provide a management plan. Landowners may prepare the plan by themselves, seek assistance from the Federal Natural Resource Conservation Service or local agricultural extension agents, or hire a private consultant to prepare the plan. The management plan is reviewed by biologists, professional land managers, and attorneys and must be approved by the Florida Department of Environmental Protection or other state agencies participating in the acquisition or monitoring.

Once the easement is purchased, a state agency — often the Florida Fish and Wildlife Conservation Commission or the Division of Forestry — monitors the easement for compliance. Monitoring is usually a straightforward matter of inspecting the site periodically to affirm that activities prohibited under the easement are not taking place. TNC has provided technical assistance to help state land-management agencies develop monitoring protocols.

Economic Incentives

Several landowners in the region surrounding the Florida Panther Corridor have expressed the opinion that they could make more money selling to developers or other business interests than to the state of Florida for conservation purposes. In many parts of Florida this may be true. Developers have been known to pay above appraised value to acquire a particularly desirable tract for development, while statutes generally prohibit the state from paying above appraised value.

Many landowners in the Florida Panther Corridor region, however, have chosen to sell their lands or to sell conservation easements because they have a long family history in rural Florida and would rather see their land conserved than subdivided and, potentially, developed. When working with these landowners, TNC and the state of Florida emphasize the value of conserving lands for posterity and the overall benefits the land can provide to the citizens of Florida as a source of water, wildlife, and other natural values. In consultation with and always at the discretion of state agencies, TNC sometimes suggests the naming of a conservation area after the family that sells the land. The primary incentive for a landowner to do business with the state under the P2000 and Florida Forever programs is, however, economic. For example: (i) these programs pay at or near fair-market (i.e., appraised) value; (ii) they have an excellent track record for closing deals efficiently; and (iii) they pay in cash, usually as a single, lump sum payment.

Landowners who donate lands or easements to TNC may also receive substantial tax benefits. Donors who give an easement over their land as a gift to a nonprofit organization can deduct its worth from their federal income taxes for up to five years, taking as much as 30 percent annually off their gross adjusted incomes. When an owner retains ownership but grants an easement that cedes development rights for the land, the appraised value (and taxable basis) of the land, and in some cases the local property taxes, are also reduced. In addition, a donor's heirs will have a concomitantly reduced inheritance tax burden on lands under a conservation easement, which can save them from having to subdivide and sell off the inherited property — a strong negative consequence for important conservation lands. Since state funding has become more readily available for easements, most owners have now opted to sell easements. Particularly for ranchers, who tend to be land rich but cash poor (and who hold some of the best remaining conservation lands in Florida), the sale of an easement at appraised value is often an attractive option.

As discussed above, county governments may sometimes resist setting aside new state or federal conservation areas unless they are assured that lost tax revenues will be replaced from other sources. Revenues from timber sales are an incentive, and economic activities such as ecotourism may help bring people into their counties. Coupled with revenues from the sale of hunting and fishing licenses, this can help compensate the county for tax revenue it could have collected from private landholders. The amount of money local governments can directly collect from public land depends on its management status. For example, counties can receive 15 percent of timber proceeds from state forest land and 25 percent of timber proceeds from federal (i.e., U.S. Forest Service) land. In addition, Florida Forever also provides funds to replace local property tax revenues lost in qualifying rural counties. This payment in lieu of taxes provides annual payments equal to the amount of property taxes lost for ten years (Florida Department of Environmental Protection 2000).

An abandoned railroad provided an additional economic incentive for Hendry County to support conservation of the Caloosahatchee Ecoscape project. Working with the U.S. Fish and Wildlife Service, TNC helped to educate local officials about the overall economic potential and importance of the envisioned corridor and offered to provide technical support for its implementation under the Florida Division of Environmental Protection's Rails-to-Trails program. After several months of discussion, county officials agreed to support conservation of the Caloosahatchee Ecoscape because they saw trail development as a way to foster nature-based economic activities. These have been demonstrated to generate substantial revenues in other regions of Florida.

Governance

Deciding who will manage newly acquired public land is another challenge of the implementation process. As illustrated in the case of the Pinhook Swamp Corridor (box 5.4), differing expectations sometimes arise over which agency will manage newly acquired conservation land. As part of its land acquisition proposals, TNC offers management recommendations based on the landscape condition, historic uses, and conservation needs (including habitat and hydrologic restoration) of each property and on the specific expertise and management philosophies of different resource management agencies.

BOX 5.4. NEGOTIATING AND GOVERNING
THE PINHOOK SWAMP CORRIDOR

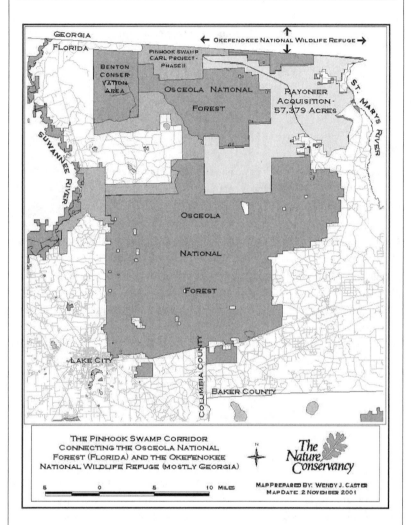

The Pinhook Swamp Corridor
Connecting the Osceola National
Forest (Florida) and the Okefenokee
National Wildlife Refuge (mostly Georgia)

The Nature Conservancy

MAP PREPARED BY: WENDY J. CASTER
MAP DATE: 2 NOVEMBER 2001

The 23,221-hectare Pinhook Swamp Corridor connects the Osceola National Forest in Baker and Columbia Counties of northeastern Florida to Georgia's Okefenokee National Wildlife Refuge. The corridor is designed to protect habitat and provide connectivity for several threatened or endangered species — including Florida black bear (*Ursus americanus floridanus*), Sandhill Crane (*Grus canadensis pratensis*), and Red-cockaded Woodpecker (*Picoides borealis*) — and it

continued

has been proposed as a possible area for reintroduction of the Florida panther. In addition, the Pinhook protects watershed resources for the city of Jacksonville and other nearby cities. The Pinhook corridor can also support recreational activities such as hunting, fishing, canoeing, and hiking. With the purchase of the final segment in April 2001, the entire protected area now connected by the corridor encompasses 50,815 hectares of mesic and wet flatwoods, cypress-bay-tupelo swamp, and floodplain forest habitat (Hilsenbeck and Caster 2000).

TNC helped the U.S. Forest Service acquire a portion of the western Pinhook area from one family during the early 1990s. An opportunity to work with the Rayonier Timber Company — which held the last and largest gap in the corridor — took several more years. When Rayonier decided it was ready to negotiate for a possible sale, TNC and the state of Florida initially recommended a phased, two-part sale so as to ease the impact on available state conservation funds. The combination of the business-savvy Rayonier and the state, eager to complete the entire long-awaited corridor, ultimately made the phased approach less than acceptable. As an outside party, TNC helped negotiate the deal by convincing both parties that a fair transaction could be concluded within an acceptable time frame. The final deal involved a single phase, fee-simple purchase costing nearly US$60 million. Seventy-five percent of the funding came from the Division of State Lands (a part of Florida's DEP) through P2000, while 25 percent was supplied by Florida's St. Johns River Water Management District.

The Pinhook Corridor provides an example of corridor implementation through opportunity, persistence, cooperation between a corporation, the government, and a third-party NGO, and enough public funding to make such a large conservation acquisition a reality. Bottom-line financial considerations probably provided the primary incentive for Rayonier to sell its Pinhook Swamp property. After buying out another company's land holdings, Rayonier may have decided to sell some of its excess holdings to raise cash or reduce corporate debt. Likewise, a convincing proposal by TNC outlining the biological significance of the Pinhook Swamp helped persuade the state lands program that it was a tremendous conservation investment. The state and the federal government will earn some revenue from the property through timber harvests and sale of hunting and fishing licenses.

The Pinhook eventually may be managed by two or three agencies. Determining how to manage newly acquired Pinhook Swamp habitat has required bargaining between the state of Florida and the Federal Bureau of Land Management (BLM). The BLM owns mineral rights underlying some state forest and other state conservation lands in other areas of Florida. At the request of BLM, the state must buy out these mineral interests in order to avoid the potential mining of these lands for sand, limerock, and/or petroleum. After prolonged negotiation (still not complete as of August 2002), the state and BLM may agree to swap land title (and management rights) to about one half of the Pinhook

continued

Swamp Corridor in exchange for the mineral rights in these other conservation areas. The U.S. Forest Service is expected to manage the western half of the corridor, contiguous with the Osceola National Forest, while the eastern half is currently managed by the Florida Division of Forestry.

In the case of the Florida Panther Corridor, governance of state lands is shared, to date, by two agencies — the Florida Fish and Wildlife Conservation Commission and the Florida Division of Forestry — based on their expertise in managing the various kinds of habitats and wildlife present, and on their expertise in applying management practices such as restoration and prescribed fire. Each agency has sought to manage about 50 percent of the Florida Panther Corridor area acquired to date.

In most cases, a single governing agency is designated to assume responsibility over a given area and true co-management (with equally shared management dollars) between government agencies is uncommon. Typically, only the lead agency receives management funding. Natural resource agencies tend to manage their holdings according to their prevailing management philosophy, even when these are contiguous with lands managed by other parties. To ensure that management plans are compatible with conservation goals, all management plans are reviewed by other state agencies, state biologists, and NGOs, under a public review process. When state and/or federal agencies disagree over the jurisdiction of a new public land, governance arrangements are typically established through negotiation. Which agency originally acquired or paid for the land and its proximity to current management units are factors of considerable significance in deciding such issues.

Private landowners continue to govern activities on their lands within the specified terms of conservation easements. Florida state agencies, or sometimes TNC, are those entities most likely to monitor the various conservation easements within the Florida Panther Corridor, preferably on an annual or semiannual basis once acquisition is complete. Monitoring involves ensuring that only the activities specifically allowed in the conservation easement are occurring on the land and that no violations of these provisions have occurred. The person or agency charged with monitoring a conservation easement preferably should be familiar with the condition of the land at the time the easement was granted, with the numerous provisions of the conservation easement, and with overall land-management practices.

The easement documentation report (EDR), provided by the Florida Department of Environmental Protection, TNC, or a third-party contractor, is an indispensable tool for monitoring conservation easements. The EDR consists of detailed text, up-to-date maps, aerial photographs showing the current condition of the property, and numerous GPS-referenced site condition photographs that provide a baseline description of the physical and biological condition of the property, including human-made structures. The EDR protects both the state's conservation interest in the land and the landowner's rights in case a dispute arises over the activities that have actually occurred on the property since the granting/sale of the easement. Analysis of recent aerial photographs, frequent flyovers, and, whenever possible, annual ground surveys by professional staff accompanied by the land manager are the most effective way to monitor conservation easements. In the experience of TNC, monitoring has usually proven to be a straightforward process, but it does require time, funding, and regular site visits. Depending on who holds the conservation easement, easement violations are reported to the appropriate entity, which then contacts the landowner and initiates action to ensure the immediate cessation of the violation and, if necessary, to restore any damage to the conservation values of the property.

Conclusions

The Nature Conservancy's efforts to establish the Florida Panther Corridor provide practical insight for corridor design and implementation. By identifying priority corridor sites and working closely with landowners, TNC is striving to make the most of existing state funds for the conservation of wide-ranging vertebrate wildlife through well-planned and designed landscape-scale conservation projects.

Corridor design is based on a landscape-level vision for providing connectivity between federally protected lands in southern Florida and suitable panther habitat in central Florida. The "ideal" corridor path is identified based on habitat values, connectivity, and the actual travel routes used by radio-collared panthers. The final configuration of the corridor and the time it takes to implement the design may depend, however, as much on opportunities to engage willing sellers as on the biology that compels the corridor's establishment. Implementation is thus proceeding through a series of independent but interrelated projects.

The Florida Panther Corridor also serves multiple purposes: it provides habitat for a variety of rare, threatened, and endangered species, protects valuable watersheds, and, in some areas, will support substantial recreation.

The multiple benefits of corridor lands have aided in attracting support from state and local governments and other funding partners.

Major stakeholders in Florida Panther Corridor implementation include private (often corporate) landowners, state funding agencies, county governments, and sometimes federal agencies. Carefully crafted proposals, including data on panther movement, have helped secure government prioritization and funding. Various economic incentives — including payment in lieu of taxes and recreation development for county governments, and efficient and fair real estate transactions for private landowners — help to engage these stakeholders.

Land tenure arrangements and governance in the Florida Panther Corridor will range from private ownership and management under conservation easements to management of new state (or, perhaps, federal) conservation lands by various government agencies. With each proposal, TNC makes recommendations regarding governance based on the particular expertise of natural resource agencies for dealing with the conservation needs of the project. In some cases, multiple agencies may manage tracts, but true co-management (with equally shared dollars) is uncommon.

Case 4: The Y2Y Corridor in the U.S. and Canadian Rocky Mountains[11]

Summary

The Yellowstone to Yukon Conservation Initiative, or Y2Y, is a long-term effort to protect an intact habitat corridor spanning the Rocky Mountains from Yellowstone National Park, USA, to the Canadian Yukon. The ultimate goal of Y2Y is to maintain and restore a network of core wilderness areas, buffer zones, and multiple-use management areas to preserve biodiversity and ecosystem integrity, including movement of multiple far-ranging species. Y2Y is an umbrella organization that provides research, funding, technical support, outreach, and information-sharing services to other NGOs, citizen groups, and government agencies working to implement conservation projects that contribute to the overall corridor vision. This case study is particularly useful for examining corridor design and implementation.

Y2Y is using GIS technology to create a regionwide map of priority core areas, buffer zones, and corridors based on the requirements of a suite of terrestrial and aquatic focal species. So far, Y2Y has completed a grizzly bear

habitat–suitability study, which identifies optimal pathways for bear movement. In cooperation with numerous GIS labs in Canada and the United
States, Y2Y is compiling a seamless database that will incorporate data sets
ranging from carnivore distribution and net primary productivity to land
uses and public attitudes. A map incorporating these data layers will provide a template for prioritizing local conservation efforts in the context of
the landscape-level conservation vision.

The Y2Y Initiative envisions that a landscape-level corridor will be implemented through multiple, smaller conservation projects. Initiatives in western
Canada illustrate the kind of local efforts that the Y2Y Initiative hopes to
support. An example from Canmore, Alberta, illustrates community-based efforts to utilize state-of-the-art principles in designing conservation landscapes.
Two additional projects — the Muskwa–Kechika Management Area of British
Columbia (BC) and community visioning in Revelstoke — are particularly
useful for understanding stakeholder engagement, and they illustrate the use
of consensus-based conservation planning. The examples also address issues
of land tenure, economic incentives, and governance likely to shape corridor
implementation throughout the Y2Y region.

Background

Initiated by a concerned environmental lawyer and members of the Wildlands Project in the early 1990s, the Yellowstone to Yukon Conservation
Initiative seeks to maintain habitat connectivity for wildlife inhabiting the
Rocky Mountains of the United States and Canada (fig. 5.5). Outreach Coordinator Jeff Gailus describes Y2Y as a service organization helping to coordinate the efforts of local conservation groups that actually implement
conservation on the ground. Y2Y provides a larger vision and context for
local efforts and contributes scientific research to assess wildlife needs and
guide regionwide conservation planning. It also provides funding to local
groups through small grants and assists groups in finding other funds. In
addition, Y2Y hosts workshops on capacity building and facilitates information exchange and networking to enhance cooperation and minimize
duplication of effort between conservation groups. In these ways Y2Y advocates for "the long-term future of wildlife, wildlands, and sustainable communities," but as an organization it rarely takes a stand on specific campaigns. A board of directors, composed of representatives from network
groups, charts the course of Y2Y.

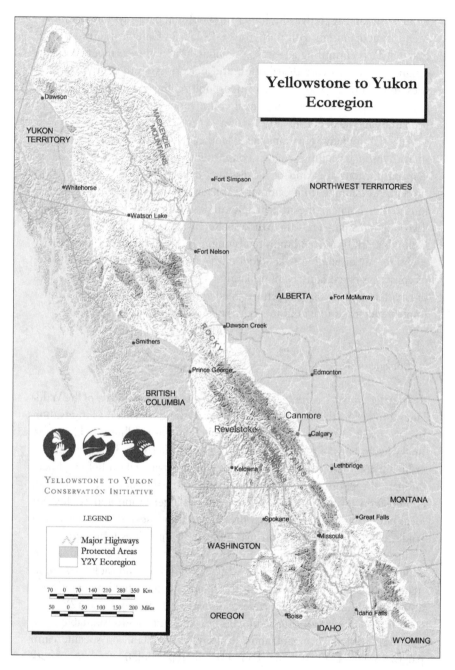

FIGURE 5.5

The Yellowstone to Yukon ecoregion. Courtesy of Yellowstone to Yukon Conservation Initiative (http://www.y2y.net).

The greater Y2Y Corridor encompasses about 1,200,000 square kilometers of Rocky Mountain habitat above 1,050 meters elevation (Willcox 1998). It stretches 3,200 kilometers long and is 200–400 kilometers wide, extending from west-central Wyoming (42° N) north to the Canadian Yukon (66° N) (Gadd 1998). The Y2Y Corridor includes parts of British Columbia, Alberta, Yukon, and the Northwest Territories of Canada, and parts of Washington, Oregon, Idaho, Wyoming, and Montana in the United States (fig. 5.5). Common ecological, hydrological, and geological processes, species assemblages, and conservation threats unify this vast region (Willcox 1998).

The northern half of the Y2Y Corridor includes large expanses of habitat where native species assemblages and ecological processes remain intact. In the Northern Rockies, design thus focuses on preserving connectivity and planning where human development should take place. The southern, U.S., portion of the corridor region is much more disturbed, including rapidly growing towns and heavily visited national parks. In the Southern Rockies, implementation must sometimes focus on reestablishing movement and habitat corridors to restore connectivity for wildlife (Marcy Mahr, personal communication).

In terms of habitat composition, about 59 percent of the Y2Y Corridor is forested, primarily with mixed deciduous forest (16 percent). Major vegetative communities include grassland, dry range, and shrubland, with pine forest in the south. Wet shrubland, tundra, mixed deciduous forest, spruce and fir forests, and bare rock outcroppings dominate in the north and at high elevations. In addition, cedar and hemlock coniferous forests grow in pockets of maritime climate (Merrill and Mattson 1998a). The corridor also incorporates extensive riparian and aquatic habitat, including 320 major watersheds (http://www.y2y.net/).

At least 70 mammalian species inhabit the Y2Y Corridor. Potentially viable populations of several far-ranging carnivores, including grizzly bears (*Ursus arctos horribilis*), wolves (*Canis lupus*), and wolverines (*Gulo gulo*) — are now restricted to the less developed, northern portion. At least 275 bird species breed in the Y2Y region, including species dependent on mature, interior forest habitat such as the Northern Goshawk (*Accipiter gentilis*) and Barred Owl (*Strix varia varia*), and riparian-dependent species, including the American Dipper (*Cinclus mexicanus*) and Harlequin Duck (*Histrionicus histrionicus*). Millions of migrating raptors and passerines also use the Y2Y as a stopover point (Holroyd 1998). The corridor region's waterways support at least 118 species of fish, including popular commercial species such as Chinook salmon (*Oncorhynchus tshawytscha*) and endangered, endemic species such as west slope cutthroat trout *(Oncorhynchus clarki lewisi)* (Mayhood et al. 1998).

About 10 percent of the Y2Y corridor is currently protected within eleven national parks, numerous wilderness areas, and regional parks (Merrill and Mattson 1998a, Webster 1999). Grassland, mixed forest, and cedar/hemlock forest ecosystems are not, however, represented in protected areas in proportion to their natural abundance (Merrill and Mattson 1998a). Y2Y biologists believe that none of the existing reserves is large enough to support populations of far-ranging species — such as grizzly bears, wolves, and wolverines — over the long term. Maintaining connectivity between protected areas is essential for conserving a viable Rocky Mountain ecosystem (Merrill and Matson 1998b).

The human inhabitants of the Y2Y Corridor region include thirty-one Native American, or First Nation, groups. Research suggests that what early European explorers perceived as "pristine" Rocky Mountain ecosystems were already shaped by human activity. Native practices such as burning and cultivation of medicinal plants may have helped to maintain biodiversity in the region. Many First Nations have left their traditional homelands within the last 150 years, but some communities in the Canadian Rockies maintained a fur-trapping era lifestyle (circa 1800) until the late 1970s. As remaining subsistence communities continue to age and shrink, traditional lifestyles and ecological knowledge are being lost (Pengelly and White 1998, Reeves 1998). For this reason, Y2Y is interested in incorporating indigenous knowledge into conservation planning.

The modern economy of the Y2Y region was traditionally resource-based, but it has diversified greatly in the last decade. Throughout the 1980s many Rocky Mountain communities experienced economic depression and population decline as resource-based industries such as agriculture, logging, and mining slowed down. The 1990s witnessed rapid "amenity-based growth" as businesses and retirees began moving into the region, drawn largely by the beautiful natural scenery and the rural lifestyle. Since the 1990s, the largest source of economic growth has been "nonlabor" income from the pensions and investments of retirees.[12] Other major areas of growth include services and professional industries. Average income increased during 1990s even as employment in traditional extractive industries remained stable or declined (Rasker and Alexander 1998). Given these trends, conservation must balance habitat preservation with pressure for both continued natural resource extraction and urban development.

Large mammals in the Y2Y Corridor are threatened by habitat destruction and fragmentation and by displacement and mortality caused by human activities. Extermination of large carnivores such as bears and wolves by humans is a major problem. Bird species are likewise threatened by habitat loss, particularly riparian-dependent species.[13] Habitat loss and impoundment, overfishing, competition from introduced species, and water pollution threaten fish and aquatic communities (Mayhood et al. 1998).

Human pressure on native species will likely increase in the future as a result of rapid population growth and development in the Y2Y region. Major sources of human disturbance include:

- oil and gas development;
- industrial forestry;
- suburban sprawl;
- fire suppression;
- agriculture (particularly livestock grazing);
- mining;
- outdoor recreation, including heavy visitation to national parks and activities such as off-road vehicle driving, hiking, kayaking, horseback riding, hunting, fishing, and camping; and
- toxic pollution associated with industrial development and increased traffic.

Documented impacts of these activities include habitat loss and fragmentation, alteration of natural hydrological cycles, simplified vegetative communities, soil damage, sedimentation of rivers, groundwater contamination, altered fire regimes, and wildlife mortality and displacement. Studies in the Y2Y region estimating the impacts of different human activities according to kilometer of linear disturbance generated (e.g., roads, pipelines) suggest that oil and gas development and forestry are the most damaging activities (Sawyer and Mayhood 1998). Based on these threats, proponents of the Y2Y Corridor argue that the pace and scale of growth must be controlled to ensure both ecological and economic health for the region (Sawyer 1998).

Design and Implementation

The Y2Y Initiative contributes to corridor design primarily through research. Initial research has focused on grizzly bears and other potential umbrella species. Y2Y scientists intend to integrate data on a suite of focal species — including several carnivores, ungulates, birds, and fish — to develop a tool for identifying conservation priorities throughout the landscape. This tool will be used to pinpoint a system of core protection areas, smaller-scale corridors, and transition zones. Longer-term research priorities include mapping of human attitudes, modeling of potential land-use changes, and research on potential climate change impacts in the Y2Y region (Mahr et al. 1999).

To date, Y2Y has completed a habitat-suitability map to identify potential core areas and corridors for grizzly bears in the Canadian Rocky Mountains, using indicators of habitat quality (vegetative cover), habitat heterogeneity (forest/grassland interface), and human disturbance (road density). The grizzly is an attractive focal animal because it is wide-ranging and particularly sensitive to development, and because grizzly life history traits are the best understood of the carnivores in the region. In addition, grizzly habitat correlates with that of several other species of conservation interest, including wolves, wolverines, lynx (*Lynx rufus*), Northern Goshawks, and Black-backed Woodpeckers (*Picoides arcticus*) (Herrero 1998). The model identifies pathways that could offer the best chance of survival for dispersing bears, but it cannot predict or ensure that these paths will actually be used. Such pathways tend to follow mountainous terrain and tree cover, and their length ranges from under 200 to almost 400 kilometers (Walker and Craighead 1998).

Y2Y research is moving beyond the single focal species approach in the hope of developing more reliable indicators of habitat connectivity and biodiversity preservation. A study by the aquatic research group of Y2Y highlights a potential danger of planning based on single umbrella species (Wuethrich 2000). The study found that reserves and corridors designed to protect grizzly bears in Swan Valley, Montana, failed to protect habitat for endangered bull trout (*Salvelinus confluentus*). Furthermore, protecting the grizzly habitat could funnel more intensive human activity into key areas for aquatic conservation. The study concluded that corridor design should consider such potential conflicts before implementation begins. Communities and governments might become less sensitive to conservation planning, they felt, if advocates launch repeated initiatives to expand or alter conservation boundaries as knowledge accumulates regarding the needs of additional species.

Funded by a grant from the U.S. Federal Data Committee, Y2Y is working with a consortium of more than twenty GIS labs at private companies, universities, state and federal agencies, and NGOs to compile and integrate priority data layers recommended by its science advisors (box 5.5). These data layers will be used to build a landscape-level conservation design that includes the distributions and habitat requirements of multiple focal species and other ecological features of interest in relation to human impacts, including current and future land uses. The first project of this consortium, completed in late 2001, drew largely on preexisting data sets to build a model of how human activities such as road building impact habitat connectivity for wildlife in the Northern Rockies. The model, known as a cumulative effects analysis, is available on the Y2Y website (http://www.y2y.net/).

BOX 5.5. PRIORITY DATA LAYERS FOR A GIS-BASED
CONSERVATION AREA DESIGN

Some of the highest-priority data layers identified by conservation biologists for the Y2Y Corridor region include:

Carnivore habitat suitability, including prey density;
Measure of remoteness, including road density, human density, and disturbance;
Land cover and vegetation, including net primary productivity;
Terrain complexity;
Present and near-term land use (including attitudes of key interest groups).

— Source: Mahr et al. 1999

The Y2Y database is currently being used to explore such key questions as what constitutes connectivity for avian species, and what kind of connectivity is desirable in aquatic systems. Once key data layers, such as species distributions, habitat types, hydrological and topographical features, and land uses, are combined, it will be possible to monitor key processes and compare conservation options. Through cumulative effects analysis and other GIS techniques, for example, the Y2Y database could be used to identify population sources and sinks for focal species, or to follow the distribution and spread of exotic species to identify sources of introduction. Likewise, different management strategies — such as protection versus multiple use and private versus public ownership — could be compared to find the best conservation investment.

Although GIS technology is allowing Y2Y to rapidly compile and synthesize large amounts of data, it is not immune to unexpected challenges related to international corridor planning. For instance, the different data-sharing laws in the United States and Canada differ. In the United States, considerable amounts of data gathered by government resource agencies are available to the public free of charge. In Canada, however, public agencies generally discourage data sharing and charge users for access. In addition, private firms control much of the data Y2Y needs from Canada and charge hefty fees for sharing. This presents a major problem for Y2Y's land-use planning strategy, which depends on making the database, and ultimately the entire conservation area design, available to partner organizations and to the public at large. Y2Y and its GIS partners have formed a new legal entity, the Rocky Mountain Data Consortium, which, it is hoped, will be able to purchase rights to Canadian government data collectively. If successful, this approach could free the Data Consortium to share data among its various members, but it still would not solve the problem of how to post the data for public use.

Y2Y envisions a landscape-level corridor comprised of multiple smaller conservation initiatives. Such initiatives have begun independent of the Y2Y Initiative through the efforts of local groups. A publication authored by Jeff Gailus (2000) — entitled *Bringing Conservation Home: Caring for Land, Economies, and Communities in Western Canada* and published jointly by Y2Y and the Sonoran Institute — presents case studies illustrating how community-based activities provide a foundation for the Y2Y Corridor.

Example 1: Corridors in the Bow Valley, Alberta   The town of Canmore, Alberta, has begun designating wildlife corridors in the Bow Valley as part of its municipal growth management plan (see location in fig. 5.5). The Bow Valley provides a critical east–west link for wildlife such as caribou and bear, which complete seasonal migrations along adjacent north–south running valleys (Gailus 2000). A recent report commissioned by Parks Canada and local conservation organizations suggests that the southern Canmore area is particularly important for wildlife (BCEAG 1998).

The southern Canmore area is a strip of land about 10 kilometers long and 1.5 kilometers wide on the southern edge of the Bow Valley that links two protected areas (Wind Valley Natural Area and Canmore Nordic Centre Provincial Park). Because major highways and a railroad disrupt other movement routes, this narrow strip provides one of the last intact linkages for wildlife moving from Banff National Park into the greater Y2Y Corridor. In addition, obstruction of wildlife movement through Bow Valley could sever the northern and southern portions of the greater Y2Y Corridor (Herrero 2000). Concern for preserving wildlife movement has led to the establishment of a new protected area as well as several corridor projects in the southern Canmore area.

Once a small coal mining town, Canmore has experienced rapid population growth since hosting the 1988 Winter Olympics. Between 1996 and 2001, population increased by over 50 percent, with annual growth rates reaching up to 10 percent. Today Canmore's population is about 10,000 and is projected to nearly triple by 2015 (Herrero 2000). Pressure for increased residential development, and especially large-scale resort development, is very strong. By the early 1990s, residents had become concerned about how growth was affecting their communities and the local environment.

Following a contentious and only partially successful court battle to stop a major tourist development from blocking a wildlife corridor in an adjacent valley, local conservation groups and sympathetic city officials in the Bow Valley formed a task force to seek more proactive solutions. The task force produced maps to illustrate how development could disrupt wildlife corridors and began a campaign to involve government agencies at the local, provincial, and national level in developing collaborative solutions.

The activities of the local task force inspired the provincial government of Alberta to create a Bow Corridor Ecosystem Advisory Group (BCEAG). This is composed of senior planners from relevant government agencies[14] that research corridor issues and coordinate regional planning. In 1998, BCEAG published *Wildlife Corridor and Habitat Patch Guidelines for the Bow Valley*, which provides specific criteria for the design of multispecies wildlife corridors (box 5.6).

BOX 5.6. BCEAG GUIDELINES FOR WILDLIFE CORRIDORS AND
HABITAT PATCHES IN THE BOW VALLEY

The Bow Corridor Ecosystem Advisory Group (BCEAG) provides detailed guidelines to help planners (primarily developers) in the Canmore region design wildlife corridors in compliance with the town's growth-management plan. The guidelines provide formulae for relating variables of length, width, shape, topography, vegetative cover, and adjacent land uses in corridor and patch design. This approach is based on studies of wolves and bears in Banff National Park and in surrounding ecosystems, studies of corridor use by cougars in California, and conservation biology theory.

continued

Definitions: BCEAG guidelines differentiate types of corridors and habitat patches. *Primary corridors* are the major corridors linking protected areas over long distances, support movement of multiple species such as large carnivores and others sensitive to human disturbance, and must be at least 350 meters wide. *Secondary corridors* link habitat patches over shorter distances for smaller and/or more human-tolerant species such as elk and must be at least 250 meters wide. *Local habitat patches* provide food and rest for animals passing through the corridor network to larger *regional patches*, which are protected areas extensive enough to support large carnivores for a limited time.

Conditional factors: The minimum width requirements vary according to parameters such as topography, slope, and vegetative cover. For example, corridors with a length greater than 1 kilometer or vegetation cover less than 40 percent may require increased width. Conversely, topographical features — such as gullies that channel wildlife movement and provide protection from human disturbance — may reduce the width needed, while a grade steeper than 25 percent is considered nonfunctional for animal movement.

Shape: BCEAG recommendations address appropriate shapes for patches and corridors. Corridors should be as wide and straight as possible, while patches should maximize the ratio of surface area to perimeter. Design should avoid doglegs and peninsulas, which may slow passage through the corridor network or lead wildlife into hazardous edge areas. Multiple large corridors should connect regional habitat patches to increase the chances that wildlife will use them and to guard against connectivity loss should one corridor be disrupted.

Human disturbance: BCEAG provides guidelines for minimizing human impacts on corridors and habitat patches. It specifies limited activities that corridors may accommodate and recommends setbacks for different types of development adjacent to corridors. Guidelines also describe an approval process that requires developers to perform wildlife impact assessments, map corridors according to standardized procedures, and monitor use by wildlife.

— Source: BCEAG 1998

Humans and wildlife tend to favor the same habitat in the Bow Valley, which presents a major problem for the establishment of additional corridors. Flat valley bottoms form natural corridors used by migrating animals and provide montane habitat used as a winter refuge by elk, deer, bighorn sheep, and carnivore species. These flat areas also offer the best locations for building resorts. The Bow Valley Provincial Park mostly occupies the southern slopes of the Bow Valley because tourism companies own much of the valley floor (Herrero 2000).

Three Sisters Resorts, which plans to develop a major tourist lodge on the floor of Bow Valley, has proposed several small corridors designed before the BCEAG guidelines were published. Problems with these proposed corridors highlight the importance of selecting appropriate data for corridor design. Resort contractors used vegetation cover specifications developed for two relatively human tolerant ungulate species (deer and elk) on nonwinter ranges in Oregon and Washington. As proposed, the corridors are too narrow to provide sufficient winter cover or to provide habitat for carnivores and other species that tend to avoid humans (Herrero 2000).

Canmore has so far developed several new land-use designations to preserve corridors, but it has not yet established the final configuration for any corridor within the Bow Valley. BCEAG and the municipal government can "proactively encourage" adherence to BCEAG guidelines, but compliance can be enforced only if the municipality adopts specific legislation (BCEAG 1998). Whether viable corridors are ultimately established in the Bow Valley will depend on the ability of an advisory group without legal power to enforce its decisions.

Stakeholder Engagement

As an umbrella organization, the Y2Y Initiative helps local groups identify and involve stakeholders in conservation projects that contribute to the overall corridor vision. The outreach program of Y2Y works to gain support from stakeholders, who may officially sign on to the Y2Y network or express tacit support for the Y2Y Initiative. The mayor of one Canadian town (Invermere) has publicly endorsed the Y2Y Corridor. The provincial governments of British Columbia and Alberta have expressed support in principle but are not yet formal network members. In addition, the U.S. Park Service and Parks Canada have signed memoranda of understanding with Y2Y.

To date, more than 160 groups representing almost one million conservation-minded individuals have joined the Y2Y network. Major stakeholder groups include local governments, citizens, sportsmen (hunters, anglers), ranchers, and First Nations. For example, Y2Y appeals to hunters and anglers as conservation allies and stewards of the environment. Sportsmen could benefit from new hunting and fishing opportunities in some protected areas. Two hunters' groups (Orion: The Hunter's Institute and Montana Wildlife Federation) have joined Y2Y.

Many First Nation groups maintain traditional, cultural, and religious ties to the Y2Y region and some still rely on natural resources for subsistence.

Y2Y invites collaboration from native groups to ensure that conservation planning respects cultural values and has made the inclusion of traditional ecological knowledge an explicit goal of its conservation planning. To date, the Blood Tribe of Canada has endorsed the Y2Y Corridor.

Ranchers are among the most skeptical but also most influential stake-holders in the Y2Y region, and rancher-oriented groups such as the American Land Rights Association have been vocal critics (King and Reibstein 1997). To attract ranchers, Y2Y focuses on "a common goal of open space." Y2Y is also encouraging governments and land trusts to recognize the conservation value of range land through economic incentives such as breaks on inheritance and income tax for conservation easements.

For local communities, conservation planning can bring greater local control over development and generate new economic activity through ecotourism and sustainable resource extraction. Two further examples illustrate how engagement of local stakeholders is shaping community-based conservation initiatives in the Y2Y Corridor region (Gailus 2000).

Example 2: The Muskwa–Kechika Management Area, British Columbia
In the Muskwa–Kechika region of the Northern Rockies of British Columbia (BC), conservationists, resource industries, and multiple local governments used a consensus-building approach to establish a four million hectare multiuse Muskwa–Kechika Management Area (MKMA) that is designed to balance wildlife conservation with sustainable economic activity. If successfully implemented, MKMA will encompass about 10 percent of the Y2Y Corridor (Gailus 2000).

MKMA is located near the northern border of British Columbia and is defined by the watersheds of the Muskwa and the Kechika Rivers. These watersheds connect extensive undeveloped wildlife habitat across altitudinal, north–south and east–west gradients. Characterized as "the Serengeti of the North," the MKMA region supports healthy populations of caribou (*Rangifer tarandus*), elk (*Cervus elaphus*), mountain goats (*Oreamnos americanus*), Stone's sheep (*Ovis dalli stonei*), the only Plains bison (*Bison bison bison*) population in British Columbia, and a complete carnivore community including black bears (*Ursus americanus*), grizzly bears, wolves, wolverines, coyotes (*Canis latrans*), and mountain lions (*Felis concolor*).

The region remains sparsely populated by humans (less than one person per kilometer, according to BC Stats 2001), but by the early 1990s local environmentalists — including trappers, hunters, and backcountry guides —

became concerned about the impact of extensive road building by extractive industries. Roads fragment wildlife habitat and open remote wilderness to anyone with an automobile. Leaders from two local environmental groups organized a coalition of like-minded groups to compose a mission statement calling for sustainable land-use planning in the region. They publicized their campaign by leading groups of biologists, policymakers, and reporters on backcountry tours. Publicity generated by these activities led the BC government to limit new road building for two years so that a management plan could be developed.

In the early 1990s, the BC government set up a resource-management planning process that involved local communities throughout the province. Each plan is developed through a consensus process by a so-called planning table that includes representatives of all local interest groups. In the MKMA, which spanned two management areas (Fort St. John and Fort Nelson), planning brought together diverse interest groups — including representatives from forestry, oil and gas industries (the largest employers in the area[15]), conservationists, tour operators, and officials from national, provincial, and multiple local governments. To help all participants better understand the planning area, representatives from Fort St. John who had never seen the remote Northern Rockies were given an aerial tour. This exposure allowed participants to witness the impact of oil and gas development and compare developed areas to large tracts of wilderness that could still be preserved.

After four years of negotiation, each table approved a management plan. These plans envision a network of core protected areas including extensive old growth forest, mountain lakes, and wetlands. Core areas are surrounded and connected by transition areas and special management zones where some "ecologically sensitive" logging, mining, and oil and gas operations are allowed. As originally proposed, MKMA will encompass about 4.4 million hectares. The plans create eleven new protected areas encompassing over 1 million hectares, and they designate more than 3 million hectares for special management. In June 2001, MKMA was expanded to include the neighboring Mackenzie management plan, bringing its total area to 6 million hectares (Johnstone 2001).

An advisory board comprised of diverse interest groups was established to review the implementation of the management plans in the MKMA. The board has no authority to interpret or enforce pertinent land-use regulations, but it reports to the BC Ministry of Sustainable Resource Management and oversees a US$1.3 million trust fund set up to finance further planning and other projects.

The consensus-building process that led to establishment of MKMA was difficult and imperfect. The process attempted to involve every stakeholder group with a vested interest in land-use planning. The oil and gas industry proved to be a leader in fostering compromise, and as of 2000 it was the only party besides the BC government to contribute to the trust fund mentioned above. The forestry industry was also able to compromise, sacrificing access to some core areas to ensure access to zones designated for forestry in each management plan. The mining industry, on the other hand, eventually opted out of the process. Even after obtaining concessions from the other parties, the BC and Yukon Chambers of Mines ultimately dropped out of negotiations.[16]

Likewise, First Nation groups declined to participate in the planning process because they felt it would compromise their ongoing negotiations of land claims with the BC government. The planning tables kept First Nations informed of their progress, and one tribe has since taken several seats on the Muskwa–Kechika advisory board. Some members of the advisory board who did not participate in the negotiation process may have a misguided view of the area, because they refer to it as "one big park."

As in several of the cases reviewed in this book, establishment of the MKMA began with the initiative of a few dedicated local people. Government support was also crucial for MKMA development. The BC government initiated a consensus-based planning process and then legislated the funding and protection to implement resulting plans. A consensus process involving representatives from major interest groups succeeded in producing a plan agreed to by most parties. Implementation will be a long-term process dependent on continued consensus-building and financial support from both the provincial government and interest groups.

Example 3: Community Visioning and Conservation in Revelstoke, British Columbia   This example illustrates how a community visioning process is helping to promote conservation in the Y2Y Corridor. A town of 8,000 located in southeast British Columbia (see fig. 5.5), Revelstoke fell into economic depression during the 1980s as mines closed down, timber prices fell, and major dam and other infrastructure projects were completed. In the early 1990s, Revelstoke's mayor and city officials initiated a visioning process to involve diverse interest groups in planning for the community's future. This process led to a vision statement that incorporated goals of both economic progress and environmental stewardship.

Based on the vision statement, the Revelstoke government has taken concrete actions to diversify the local economy and improve stewardship of natural resources. Following approval by community referendum, Revelstoke purchased a nearby industrial timber holding and started a community-owned forestry corporation in 1992 (while the visioning process was still under way). The Y2Y Initiative believes that this enterprise effectively maintains a mixture of values on harvested land, including diverse recreational opportunities for local people and important habitat for mountain caribou and grizzly bears. Profits are used to fund research and development for sustainable forestry practices and to help local landowners implement better practices, including value-added timber processing.

The Revelstoke community-based Bear Awareness Program represents an innovative way of helping humans and wildlife coexist along corridor routes. With support from the BC Ministry of the Environment, Lands, and Parks, the town hired a program coordinator to run education programs and operate a Bear Aware Hotline as an alternative to calling animal control. The most difficult part of the program was getting residents to implement bear-proofing measures on their properties. Management-related bear kills were reduced from thirty-three in 1993 to just four in 1998.

The examples of MKMA and Revelstoke illustrate how strong local leadership and consensus-based decision making can promote conservation of the Y2Y Corridor. Other communities and local governments, however, remain skeptical and even fearful of Y2Y's mission. Some elected officials have pledged not to support the Y2Y Corridor because they believe it will entail stricter land-use regulations. Under pressure from a few communities heavily reliant on natural resource extraction, the Federation of Alberta Municipalities and the Federation of Canadian Municipalities passed anti-Y2Y resolutions. Y2Y has yet to meet with these stakeholders face-to-face, however, and it anticipates that doing so may help to reduce the impact of an anti-Y2Y campaign funded largely by resource industry groups. The Forest Alliance of British Columbia, for example, published a review suggesting that Y2Y could cost the region 80,000 jobs (Webster 1999).

An epic hike through the Y2Y Corridor region by Canadian biologist and park ranger Karsten Heuer has proven an excellent public outreach strategy. In 1999–2000, Heuer traveled 5,400 kilometers from Yellowstone National Park, Wyoming, to Watson Lake in the Yukon Territory, stopping to show slides and talk about Y2Y in one hundred local communities. His observations confirmed that the Y2Y corridor remains largely intact and that grizzly bears and other species of conservation interest are using it (Gailus 2000, Webster 1999). The hike received wide media coverage in Canada and the United States. Perhaps more important, Heuer was able to speak with many people in remote communities that Y2Y might not otherwise have been able to reach. While the hike has not generated a lot of official support from local communities, it may have helped neutralize opposition and divert an initial backlash against Y2Y by "putting a face on conservation." Initial contacts made through the hike may also provide a springboard for further outreach in these communities (J. Gailus, personal communication).

Land Tenure

There is no single solution for resolving land tenure issues in the Y2Y region. Corridor networks will need to incorporate both public and private lands through mechanisms ranging from fee-simple acquisition to conservation easements to informal agreements with sympathetic landowners. In the examples reviewed, conservation has entailed changing the management status of public land (MKMA and Bow Valley), purchasing private land for community management (Revelstoke), and providing guidelines for private landowners (Bow Valley).

Economic Incentives and Regulations

Conservation in the Y2Y region will need to incorporate incentives to foster cooperation, particularly with large industries and private landowners.

Obtaining buy-in from industrial interests. The case of MKMA suggests that resource-based industries and developers may be willing to participate in conservation planning because they have a clear economic stake in land-use decisions. Industries recognized the necessity of compromise with conservation objectives and were willing to negotiate to secure future access to natural resources and/or land. They also were more willing to dedicate resources to conservation when government provides a complementary conservation investment. In some areas that are both critical for wildlife and ideal for development (such as the valley floor in the southern Canmore area), however, additional incentives may be required to secure corridor land.

Obtaining buy-in from ranchers. Throughout much of the Y2Y Corridor region, conservation will entail dealing with private landowners, especially ranchers. Y2Y and the Wildlands Project have identified several types of economic incentives that may be employed to foster rancher participation in the future:

- breaks on income and inheritance tax on land with conservation easements,
- monetary compensation for ranchers who lose livestock to predators,
- voluntary retirement of grazing permits,
- replacement of cattle grazing with elk trophy hunting, and
- loans for ecotourism (e.g., bird or wolf watching).

Regulations on resource use. Finally, Y2Y literature suggests that public education and government regulations, in addition to economic incentives, will play a key role in protecting Y2Y ecosystems. Such measures could include:

- education and nonlethal bear control programs to reduce human-bear conflict (as illustrated by Revelstoke's Bear Aware Hotline);
- bag limits, catch and release fishing areas, and/or periodic closure of stressed fisheries;
- regulation of off-road vehicle access and other recreational activities; and
- control of access to multiple-use areas, as was implemented in MKMA.

Governance

Since the Y2Y Conservation Initiative is comprised of multiple, smaller conservation projects, methods of governance will likely vary throughout the region. The examples presented here have in common substantial help from supportive local and regional governments. Projects in other areas may not enjoy this advantage.

Governance of the Muskwa–Ketchika Management Area (MKMA) is particularly challenging because it involves two provincial jurisdictions. Two separate regional planning tables establish guidelines for governing the area. MKMA land remains under public ownership, but implementation is overseen by an advisory board made up of stakeholder representatives. In Canmore, corridors are protected through the activities of at least three different governing bodies: the provincial government, a citizen committee for growth management, and a government-appointed corridor advisory group. Finally, in Revelstoke the municipal government has taken responsibility for improving environmental stewardship through both the community forestry program and the Bear Aware Hotline.

Conclusions

The Y2Y Conservation Initiative highlights the use of multispecies corridor design. Research sponsored by the Y2Y Initiative suggests that corridors designed for just one species or one guild (such as terrestrial carnivores) may not adequately protect other elements of biodiversity. Y2Y is using state-of-the-art GIS technology involving collaboration with multiple GIS labs to assemble a comprehensive database for conservation area design.

The Y2Y Initiative also illustrates an approach whereby a landscape-level corridor is implemented through smaller conservation projects at the local level. Y2Y operates as an umbrella organization to promote landscape-level conservation planning in the northern Rocky Mountain region. It provides a landscape-level vision for conservation within the region and provides research, technical assistance, funding, and information-sharing services to local organizations and governmental agencies that actually implement conservation on-the-ground. Three independent initiatives started by local communities in western Canada illustrate the kind of efforts that Y2Y seeks to build upon.

The Y2Y Initiative and the examples reviewed above illustrate a number of trends:

- Charismatic leaders often play critical roles in mobilizing conservation efforts.
- Such mobilization often involves introducing key stakeholders and the media to the resources they hope to protect. In Muskwa–Kechika this meant providing backcountry hiking and aerial tours. In the greater Y2Y region, a hike by a Parks Canada ranger has helped introduce local communities to the idea of new land-use planning in a nonthreatening way.
- In two cases (Muskwa–Kechika and Revelstoke), a consensus-based decision process involving diverse stakeholder representatives was instrumental in establishing plans for corridors and other land-use tools de-

signed to maintain connectivity for wildlife. The consensus-building process was time-consuming and imperfect. All interest groups did not agree to the outcome of negotiations, but the decisions were accepted because they derived from a process that people considered legitimate.

- The Y2Y Initiative is spending years in formulating a comprehensive landscape design, and designing smaller-scale conservation landscapes in the Canmore area also has involved a considerable investment of resources. It is uncertain whether sufficient resources will be available to pursue this approach for some corridor projects, but Y2Y showcases what can be done.

- In the cases where corridor development has already begun at the local level (Canmore and Muskwa–Kechika), advisory bodies without legal power to enforce conservation plans are charged with overseeing implementation. The success of these initiatives will depend largely on the ability of such advisory groups to inspire action through continued consensus-building.

Case 5: Cascade–Siskiyou National Monument: A "Wildlife Highway" for the Klamath–Siskiyou Ecoregion

Summary

The Cascade–Siskiyou National Monument (CSNM) is a landscape corridor designed to protect habitat and migration and dispersal routes for wildlife moving between the Cascade and Siskiyou mountain ranges of Oregon, USA.[17] This case illustrates that stakeholder engagement may not be enough to build support for ambitious conservation initiatives, especially in a highly charged political environment. It also shows that a protected area approach, although sometimes preferable from a conservation standpoint, is not necessarily any faster or easier to implement than consensus-based, multiple-use planning. The case emphasizes the following issues.

Design and Implementation   Based on solid scientific evidence of the need to restore and maintain habitat connectivity, the CSNM was established as a protected area linking the Cascade and Siskiyou mountain ranges. In terms of design, the CSNM represents part of a larger-scale, science-based vision for the Cascade–Siskiyou ecoregion. Conservation organizations favor establishing a protected area over multiple-use management in the Cascade–Siskiyou area, because outright protection is deemed necessary to maintain the area's unique biological communities. The Federal Bureau of Land Management (BLM) is responsible for implementing protection of the CSNM through a management plan. Conservation organizations have submitted recommendations and contributed research to inform management planning.

Stakeholder Engagement   Local activists and grassroots organizations began a campaign to establish a protected area in the early 1980s. These efforts were followed by over a decade of public meetings hosted by the Bureau of Land Management. Yet eventual designation of the area as a national monument in 1999 generated strong opposition from local timber and grazing groups backed by powerful corporate interests. The debate between proponents and opponents of the monument was exacerbated by a new federal administration, which proposed further consultations regarding the CSNM designation and boundaries, leading to a series of public meetings during 2001. In the decades prior to establishment of the monument and during the 2001 review, the decisive factors were (i) solid scientific evidence for the conservation value of the CSNM, (ii) the relatively low economic returns from resource extraction in the region, and (iii) strong popular support for its establishment.

Economic Incentives   The CSNM incorporates extensive private landholdings (38 percent) not subject to rules governing monument lands. Increased protection of key private lands is needed to achieve desired connectivity through the CSNM. The World Wildlife Fund (WWF) is pursuing two main strategies for private lands: (i) helping land trusts and other local organizations prioritize and fund acquisition of land and/or conservation easements from willing sellers, and (ii) encouraging sustainable forestry practices through Forest Stewardship Council certification and eco-labeling of timber products.

Background

The Cascade–Siskiyou National Monument (CSNM) lies in the middle of the Klamath–Siskiyou ecoregion, which harbors one of the three richest temperate coniferous forests in the world[18] (fig. 5.6). Established by presidential decree in June 1999, the CSNM serves as a high-elevation land bridge connecting the Cascade and Siskiyou mountain ranges (fig. 5.6).

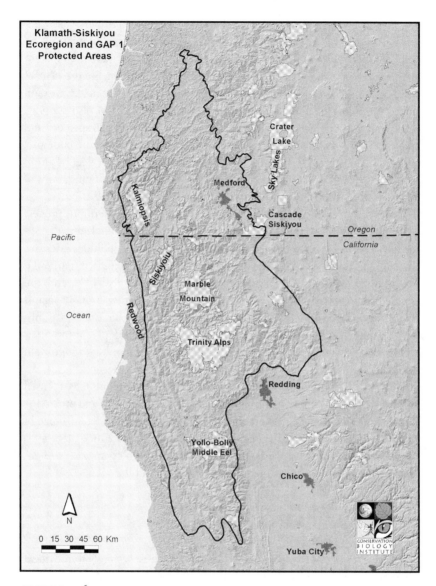

FIGURE 5.6

The Klamath–Siskiyou ecoregion. Source: Conservation Biology Institute.

Encompassing 21,044 hectares, the CSNM provides dispersal and migration routes that have been documented to function as a biological corridor. This function of facilitating dispersal and migration was a major factor in the decision to designate the area as a national monument, but the CSNM provides other conservation benefits. It protects a key ecological transition zone encompassing four major habitat types.[19] The CSNM also supports numerous rare plant communities, high butterfly diversity (120 species), and rare and endangered mollusk and fish species, many of which are endemic. Finally, the monument includes the best wintering habitat for deer in southwest Oregon and could eventually support the restoration of populations of wolves (*Canus lupus*) and other large carnivores (DellaSala 2001).

Designation of the CSNM has been a lengthy and contentious process that reflects conservation challenges prevalent throughout the Klamath–Siskiyou ecoregion. Unlike most protected areas, the CSNM includes extensive private land holdings (38 percent). Large landowner groups — in particular the Southern Oregon Timber Industries Association (SOTIA) and the Southern Oregon's Cattlemen's Association — campaigned strongly against the CSNM, suggesting that it would severely curtail private property rights. This campaign resonated in less affluent, resource-dependent communities, whereas more affluent towns and cities strongly supported monument designation.

Since the CSNM was designated, conservation organizations have focused most of their efforts on maintaining support for the monument and preserving its borders as originally proposed, and on working with the Federal Bureau of Land Management (BLM) to shape the management plan. Once this plan is approved, conservation practitioners will launch grassroots initiatives to encourage protection on private lands within and around the CSNM.

Design and Implementation

The current configuration of the CSNM was chosen largely to provide a biological corridor linking the Cascade and Siskiyou mountain ranges. The CSNM also provides substantial habitat, including centers of endemic plant and mollusk diversity, and it safeguards ecosystem services by protecting the watersheds of several major river systems. The Conservation Biology Institute (CBI)[20] provided one of the first analyses of the Klamath–Siskiyou ecoregion. Since 1996 the World Wildlife Fund (WWF) has supported protection of the CSNM as part of a landscape-level network of conservation areas spanning the Klamath–Siskiyou ecoregion (see outline of ecoregion in fig. 5.6). Box 5.7 reviews some of the techniques being used to map core areas and corridors throughout the region.

BOX 5.7. IDENTIFYING CORE AREAS AND CORRIDORS IN THE
KLAMATH – SISKIYOU ECOREGION

The CBI provided one of the first analyses of the Klamath–Siskiyou ecoregion, and it is now collaborating with WWF in designing an ecoregion-wide network of core protected areas and corridors. The objectives of the network are (i) to represent all native ecosystems, (ii) to maintain or restore viable populations of all native species, (iii) to maintain or restore ecological and evolutionary processes, and (iv) to enhance resilience to environmental change. To achieve these objectives, CBI's landscape design for the ecoregion integrates elements of special interest — such as rare species, hot spots, critical watersheds, and late seral communities — with habitat-type representation and a suite of aquatic and terrestrial focal species.

Preliminary analysis suggests that achieving these four conservation objectives will require increasing protection of more than 1,800,000 hectares (about 50 percent) of the Klamath–Siskiyou ecoregion. Timber extraction, road building, and resort development would need to be prevented in these areas. Some particularly sensitive sites (about 25 percent of the land proposed) may also require restoration and road removal. Much of the area recommended for increased conservation is public land. In privately owned areas, the strategy is to acquire land or easements from willing sellers and utilize market-based incentives such as Sustainable Forest Council certification.

— Source: Strittholt et al. 1999

Even though multiple-use management is increasingly practiced on public lands, conservation practitioners have advocated strict protection of the CSNM. They have justified this position for several reasons: (i) the economic potential for resource extraction in the CSNM is small relative to the value of vast public lands already managed for multiple uses,[21] (ii) it includes unique and fragile habitats not adequately represented in other protected areas, (iii) it supports old growth–dependent species, and (iv) it provides a critical corridor connecting the Cascade and Siskiyou mountain ranges (DellaSala 2001).

The BLM, which is responsible for implementing conservation within the CSM, released a draft management plan in mid-2002. WWF subsequently made several recommendations for monument design and management that may or may not be included in the final management plan. WWF recommendations emphasize actions to preserve and enhance habitat connectivity, particularly for old growth or late seral habitat, and to restore habitats, ecosystem processes, and endangered species. Key elements that may be addressed in the management plan include (DellaSala 2001):

- A ban on commercial logging on public lands within the monument.
- Elimination of future grazing allotments on public land and buyout of existing grazing permits from willing sellers. The executive order that created the CSNM designates funding for grazing allotment buyouts, but this initiative has generated hostility from local timber and grazing groups, who view it as an attempt by the federal government to curtail land-use rights.
- Strategic road and trail closures, including limits on motor vehicle use of BLM-designated roads. This has been a particularly contentious issue with local off-road vehicle users and property rights groups.
- Guidelines for restoring, maintaining, and monitoring old-growth forest habitat.
- Guidelines favoring the use of fire as opposed to selective thinning for forest restoration (except where needed to reduce fire risk along borders of private landholdings).
- Stream and watershed restoration (including strategic closure and re-planning of roads).
- Guidelines for conserving mollusk diversity.
- Exotic species control.

Once the management plan is finalized, WWF and local conservation organizations may pursue a variety of longer-term strategies for implementation (box 5.8). These include acquiring private inholdings from willing sellers and extending the borders of the monument. As discussed under the section on stakeholder engagement, obtaining cooperation from private landowners within and around the CSNM will be crucial for maintaining and eventually enhancing conservation value.

BOX 5.8. LONG-TERM STRATEGIES FOR ENHANCING THE
CONSERVATION VALUE OF THE CSNM

- Seek to acquire private lands in and around the CSNM, especially key habitats for mollusks and fish.

continued

- Expand the monument to boundaries that follow natural biogeographic features, including major watersheds and forested ridgelines. This would require extending the monument across the California-Oregon border. The CSNM currently stops at the border because of strong resistance to new protected areas in northern California.
- Increase protection of old-growth forest habitat on public lands around the CSNM.
- Restore historic predator-prey interactions by encouraging repatriation of large predators such as wolves, which already have begun to reappear in the region. Again, this is a contentious issue with residents who fear that the presence of large predators may endanger people and livestock.

— Source: DellaSala 2001

Stakeholder Engagement

Major interest groups involved in the CSNM designation include conservation organizations, local landowners and other Oregon residents, timber and ranching industries, and government agencies. The impetus for the CSNM designation was begun in the early 1980s by a local activist, Dave Willis. He established the Soda Mountain Wilderness Council to bring together hobby naturalists, hikers, and wilderness lovers who had strong attachments to the area and were concerned about its future. The council helped build a broader-based constituency for the CSNM at local, state, and eventually national levels. In 1990, the regional BLM office in Medford, Oregon, began work on plans for conservation-oriented management of the CSNM. By the time President Clinton designated the area a national monument in 1999, the BLM had held more than fifteen public meetings to discuss options for conserving it (Aldous 2001, Wyden 2001). Nevertheless, monument designation sparked a heated public debate that has delayed development and implementation of the management plan.

Socioeconomic studies commissioned by WWF reveal public attitudes and perceptions that complicate stakeholder engagement in the CSNM and throughout the Klamath–Siskiyou region. Even though most economic growth during the 1990s has been in service industries (250 percent increase between 1989 and 1996, compared to a greater than 50 percent decline in timber and other resource industries), many local residents believe that the regional economy remains primarily dependent on extraction-based industries. In addition, local people tend to feel that timber and other resource-based industries are already excessively controlled and resent further regulation by the federal government. This view is common throughout the American West, where the federal government controls about 60 percent of the land. The perception of external control probably increased in the final year of the Clinton administration (2000), when the federal government designated nine new national monuments, mostly in the West.

Large landowner groups — in particular the Southern Oregon Timber Industries Association (SOTIA), the Southern Oregon's Cattlemen's Association, and the Jackson County Stockmen's Association — campaigned strongly against the CSNM, suggesting that it would severely curtail private property rights. This campaign was also strongly supported by public officials linked to industry interests, in particular the Jackson County[22] Commission (Aldous 2001, Fattig 2001b). The campaign aroused fears that designation of the monument would reduce property values and income by eliminating access to timber and subsidized grazing allotments on adjacent public land. Even though the BLM stated that regulations pertaining to the monument would not apply to private lands within its borders, the campaign also raised concerns that the regulations would limit road access to their properties and that wildfires would not be aggressively controlled within the monument boundaries. Some landowners within the CSNM were also concerned that they would face public criticism for continuing legal livestock and timber operations on their properties (*Medford Mail Tribune* 2001, Edwards 2001).

This campaign resonated strongly in less affluent, resource-dependent communities, whereas more affluent towns and cities strongly supported monument designation. Economic disparity between struggling rural communities in and around the CSNM (which tend to oppose the monument) and more affluent urban centers (where support for the monument is concentrated) exacerbates the rural-urban divide over environmental issues. Many rural residents and some political leaders also view WWF and other conservation organizations as urban-based interest groups that prioritize conservation over economic prosperity. The economic disparity between rural and urban populations prevails in many parts of the Klamath–Siskiyou ecoregion, and industry-led media campaigns have often polarized local stakeholders in a jobs-versus-environment debate (DellaSala 2001).

By contrast, local interests who do not rely on timber or ranching, and those who operate inns, restaurants, and other businesses that could benefit from increased tourism, have been strong supporters of the CSNM. These interests believe that national monument status will increase their business and property values (Fattig 2001a). Many citizens and conservation NGOs at local (e.g., Friends of the Cascade–Siskiyou National Monument, the Soda Mountain Wilderness Council) and national (e.g., WWF, the Sierra Club) levels have promoted the CSNM as a way to protect wildlife and ecosystems as well as for its aesthetic, spiritual, and potential recreational values (e.g., DellaSala 2001, Stickel et al. 2001).

In 2001, a new federal administration reopened the debate over the CSNM. In March of that year, Gale Norton, the newly appointed secretary of the Department of the Interior (which houses the BLM), sent a letter to political leaders in each of the states where new national monuments had been designated under the previous administration. The letters requested feedback regarding local feelings about the monuments. In particular, it asked for any recommendations regarding impacts on private landowners, alternative uses, and possible changes in monument boundaries (Heilprin 2001).

Monument opponents seized this opportunity to reintroduce their concerns. Jackson County commissioners formed a new Natural Resource Advisory Committee headed by representatives from local government, federal officials, and the private sector and held four more public meetings during the summer of 2001 (Fattig 2001a). Opponents of the monument, including the county commissioners, recommended either eliminating the CSNM altogether or reducing its area by one-third or more to exclude most private land and limit protection to "significant features that merit monument designation" (e.g., Ressner 2001, Fattig 2001b). Monument supporters insisted that the CSNM boundaries remain unchanged, asserted that adequate public input already had been obtained, and reiterated the ecological, economic, and other benefits of monument designation (e.g., DellaSala 2001).

The results of these four public hearings were highly contentious. The four public meetings were structured so that opponents and supporters of the monument were separated, and that comments would proceed sequentially from one group to the other — thereby encouraging polarization. The Jackson County commissioners reported that 84 percent of the comments they received opposed the CSNM, whereas the Sierra Club reviewed the same comments and reported 62 percent support (Stickel et al. 2001). The commissioners refused a recount of the vote and submitted their results in a letter to Gale Norton.

Despite this polarizing process, opponents did not succeed in changing the proposed boundaries of the CSNM. Several factors appear to have been essential in maintaining the monument. First of all, conservation advocates had assembled substantial scientific evidence for the conservation value of the CSNM, including its role as a biological corridor. Second, the series of public meetings hosted by BLM during the 1990s had strengthened local support for the idea that the area within the CSNM warranted increased protection. Third, the CSNM has small potential for resource extraction compared to many other public lands in the region. Finally, and perhaps most important, grassroots activities since the early 1980s had been successful in building and maintaining strong grassroots support for the CSNM throughout the debate.

By the time the Department of the Interior proposed reconsidering monument designation, most Oregonians believed that there was overwhelming public support for the CSNM based on previous negotiations. The 2001 debate over the CSNM became the number one story for the newspaper of Ashland, the closest major city to the CSNM. Oregon's largest newspapers (e.g., Stickel et al. 2001, *Register-Guard* 2001) and various local newspapers (Singletary et al. 2001, *Ashland Daily Tidings* 2001) published editorials supporting the monument. The CSNM also received favorable coverage in major national publications, including *Time* magazine (Ressner 2001) and *National Geographic* (Mitchell 2001). All of this publicity created strong pressure for political leaders to support the CSNM. Key political leaders in Oregon — including the governor, one U.S. senator, and four out of five U.S. representatives — responded to Secretary Norton with letters supporting the monument (Kitzhaber 2001, Wyden 2001, Defazio et al. 2001).

Even though these strategies appear to have succeeded in protecting the CSNM so far, prolonged and polarized debate between local interests groups could have a detrimental impact on the long-term conservation of this monument. The need to focus on maintaining the monument's designation and boundaries has drawn the resources of government agencies and conservation organizations away from development and implementation of a management plan. The BLM was scheduled to release a draft plan by the end of 2000, but the plan has been delayed, first for review by the Department of the Interior and subsequently to await the results of studies on grazing impacts and other issues. In addition, implementation of a management plan may encounter resistance from local opposition groups still convinced that the CSNM will harm their interests. Finally, continued fear about how monument designation may impact private land use has inspired a clear-cutting spree on private lands in and around the CSNM, as landowners attempt to harvest their resources before any new regulations take effect. An interim management plan established when the CSNM received monument designation is providing some protection to public lands within the monument,[23] but accelerated extraction on private lands may destroy valuable forest habitat before buyouts or other conservation agreements can be implemented.

In the long run protection of corridors and core habitats in the Klamath–Siskiyou ecoregion will require more effective stakeholder engagement. The WWF Ashland office is devoting significant resources to public outreach and education through workshops, displays at nature centers and public events, public television programming, articles in popular magazines and newspapers, and a variety of other outlets. It is hoped that these efforts will help to build local support for the monument and other conservation initiatives critical for enhancing landscape connectivity in the ecoregion. Outreach also involves partnerships with a variety of other organizations, including (WWF 2001a):

• *Federal agencies.* Technical support and cost-sharing for U.S. Forest Service and BLM projects relevant to the CSNM, including fire restoration, grazing impact studies, and exotic species studies.
• *Conservation groups.* Technical assistance and small grants to local and regional conservation groups to support workshops, conferences, and publications.
• *Community development groups.* Support for efforts by local organizations such as the Jefferson Sustainable Development Initiative (JSDI), which helps local communities start their own conservation and sustainability initiatives throughout the Klamath–Siskiyou ecoregion.
• *Timber companies.* Promotion of sustainable management and product certification with timber companies, local mills, lumber yards, and city councils.
• *Local and regional businesses.* Encouraging establishment of sustainable business councils that promote use of certified wood and other ecologically friendly alternatives.
• *Public schools and universities.* Faculty and students from Southern Oregon University have collaborated with WWF and other conservation organizations on graduate and undergraduate research projects that benefit conservation in the CSNM. Students have mapped habitat types within the monument, and a faculty member is currently developing a case study of political factors shaping conservation in the CSNM. Graduate students have designed environmental education programs for many local schools. About thirty schools in the Klamath–Siskiyou ecoregion have begun educational programs that utilize the CSNM as an outdoor learning laboratory.

- *Native American tribes.* WWF may eventually work with other conservation organizations, such as Defenders of Wildlife and Native American tribes, to promote reintroduction of wolves, which play an important ecological role in the Klamath–Siskiyou ecoregion and also have spiritual importance for many native people.
- *Groups with expertise in urban land-use planning.* The population of the Klamath-Siskiyou ecoregion is projected to grow substantially by 2030 (by up to 300,000 people), and urban sprawl may eventually erode connectivity in lands surrounding the CSNM. Several organizations in the region support smart growth — which promotes planning to reconcile conservation and economic development. WWF lacks expertise in smart growth but will support the efforts of other organizations to reduce sprawl, such as improved zoning in areas where human development and wildlife habitat interface.

Land Tenure

The BLM will continue to control and manage public land within the monument. Regulations protecting public land within the CSNM do not pertain, however, to private lands, which occupy approximately 38 percent of the area inside the CSNM (BLM Medford). About 7 percent of this private land belongs to a single timber company, Boise Cascade, which has a poor record for land stewardship (DellaSala 2000).

WWF has begun developing a guide for local land trusts, other conservation groups, and government agencies to prioritize areas for conservation and to acquire priority private lands from willing sellers in and around the CSNM. In conjunction with this effort, WWF is raising funds for land acquisition by other organizations. As discussed in the following section, market-based incentives may also help to improve stewardship of private land in the longer term.

Economic Incentives

Market-based approaches often use scientific criteria to develop best-management practices and then introduce economic incentives to reward interested landowners for improved land stewardship. Economic incentives could be a particularly important strategy in the CSNM and other parts of the Klamath–Siskiyou ecoregion because many local communities are struggling to recover from the decline of extraction-based industries. The fact that economic incentives are strictly voluntary and nonregulatory could also be very important in the CSNM because many landowners resent government control.

As discussed above, WWF is developing plans to assist local land trusts, other conservation organizations, and government agencies in acquiring private lands from willing sellers. Since WWF is not a land-management organization, it plans to contribute primarily by providing scientific criteria and tools for identifying priority conservation areas on private land, and by raising funds for purchasing land (WWF 2001b). Having readily available funding will make it easier to purchase land from willing sellers. Conservation organizations may also work with the Oregon government to secure tax benefits as an incentive for selling land or easements for conservation (DellaSala 2001).

In the near term, certification and eco-labeling of wood products may also improve forest management on private land within and around the CSNM. An organization called Scientific Certification Systems assists timber operators in the Klamath–Siskiyou ecoregion in assessing the environmental impacts of their current operations and designing best-management practices. Landowners who manage their timber sustainably can apply for Forest Stewardship Council (FSC) certification, which opens opportunities for reaching new markets and maintaining old ones.

Conservation organizations can promote FSC certification in the CSNM in two major ways. The first is through outreach to increase consumer awareness and demand for certified wood products. Several large U.S. lumber retailers, including Home Depot and Lowes, have agreed to purchase certified wood products. Use of certification to promote sustainable forestry has grown slowly, however, due to miniscule supply and low consumer awareness of and demand for certified products. Second, conservation organizations can work with landowners to encourage and/or enable them to achieve certification. Corporate landowners (which own more than 7 percent of the private land in the CSNM) have generally shown little interest in certification. Increased consumer awareness could help to arouse such interest. Some financial assistance could be useful to smaller landowners desiring FSC certification but unable to cover its costs (DellaSala 2001).

Two other types of economic incentives may eventually be viable but have yet to be applied. One is payment to landowners for carbon sequestration services provided by intact forests. Conservation organizations such as WWF are hesitant to promote this approach in the CSNM for two reasons.

First of all, the U.S. government is not participating in international negotiations for reducing greenhouse gas emissions, thereby reducing the likelihood of government support for carbon sequestration initiatives. Second,
WWF is concerned that, if not carefully designed and monitored, carbon
sequestration credits may encourage clear-cut harvesting and conversion of
natural forests to tree farms.

Jobs in restoration could also provide an economic benefit to local communities around the CSNM. The federal government can provide funding
for restoration work through the Northwest Forest Plan. Examples of restoration activities that could provide economic benefits include thinning of
small-diameter trees (diameter less than 30 centimeters) to reduce fire hazards and riparian restoration through tree planting and obliteration of unsta-
ble roads. The environmental benefits of previous projects undertaken
through this program sometimes have been dubious. Conservation organizations could work with the BLM to ensure that projects in the CSNM are
environmentally beneficial (WWF 2001b).

Governance

The BLM will manage the CSNM as part of the National Landscape
Conservation System (NLCS), a new BLM program that applies rules that
are different from those used to govern parks and monuments managed by
the National Park Service. In general, NLCS represents a shift in BLM
management priorities from a traditional use-based emphasis to a more
conservation-oriented one. Historically, BLM lands have been managed primarily to support resource uses such as livestock grazing, mining, and logging. Resource users purchase allotments, or leases, to use these resources
on public land. Management of NLCS lands places greater emphasis on
resource conservation, including preservation of wildlife, recreation, and
range restoration. BLM lands managed for habitat conservation remain open
to relatively low-impact forms of recreation (e.g., hiking but not off-road
vehicle use) and tend to be harder to access, with few paved roads or other
visitor amenities (Mitchell 2001).

As mentioned above, the management plan will not apply to private lands
within the monument. Since almost 40 percent of the land within CSNM
is privately owned, successful cooperation between federal land managers
and private landowners will be key to the success of habitat conservation.
Some interagency cooperation will also be required: the Oregon Depart-

ment of Fish and Wildlife will continue to issue permits for hunting and fishing on monument lands (BLM Ashland, personal communication).

Conclusions

The Cascade–Siskiyou National Monument is an example of a protected-areas approach to conserving a landscape corridor. Establishing a protected area may be the best way to preserve habitat and wildlife in the corridor, but this case suggests that a protected area approach is not necessarily easier or less time-consuming to implement than consensus-based, multiuse management planning.

This case also shows that stakeholder consultation is necessary but not always sufficient to build support for ambitious conservation initiatives, especially in a highly charged political environment. Establishing the CSNM entailed long-term (more than ten years) public discussion and heated debate. Almost 40 percent of the land within the corridor is privately owned, and many local landowners who raise livestock or harvest timber are afraid that CNSM will reduce their incomes and resent what they see as a government override of local resource control. Conservation organizations have so far succeeded in maintaining protected status for the CSNM by demonstrating its conservation value and its relatively minor value for resource extraction, and by building widespread grassroots support. The discourse of the debate, however, suggests that effective compromise and understanding are still lacking among key interest groups. Without a stronger consensus and unless opposing interests are engaged to seek common ground, the future prospects for this landscape corridor are unclear.

Case 6: The Lower Kinabatangan River Corridor, Malaysia

Summary

In the Lower Kinabatangan river basin in northern Borneo, rapid expansion of oil palm plantations has reduced much of the remaining forest habitat to a thin (or sometimes nonexistent) strip of riverine forest.[24] This strip provides a tenuous corridor for the region's endangered fauna — including Asian elephants, orangutans, Sumatran rhinoceros, and proboscis monkeys. Increased flooding caused by widespread deforestation upstream, combined with more frequent incursions by elephants deprived of their natural habitat, are causing widespread damage to oil palm plantations and, ironically, provide new opportunities for conservation. Key issues in this case are the following.

Design   WWF is engaged in designing a corridor that would restore connectivity in the currently fragmented 27,000-hectare Kinabatangan Wildlife Sanctuary and link it with adjacent forest management units upstream and mangrove forests downstream. This corridor would reestablish some important migration routes, especially for elephants. Yet this Lower Kinabatangan corridor has not been formulated as part of a larger ecoregional or landscape vision, which could limit its potential contribution to conservation.

Stakeholder Engagement   Key stakeholders in the Kinabatangan include the oil palm industry, local river people, the ecotourism industry, and governmental agencies. WWF Malaysia works with each of these groups separately but, because of potentially conflicting interests and cultural values, has yet to bring all groups together around a conservation vision. It has encouraged links between the tourist companies and river people who are knowledgeable about local resources. Likewise, it has begun a dialogue with the oil palm estates about the need to conserve or reestablish forest cover adjacent to the river as a way to reduce damage by flooding and elephants. Some estates have begun pilot plantations over extremely limited areas, with technical support from governmental agencies. In a recent land-use forum, representatives from the government, business, and agriculture sectors agreed to participate in the development of a master plan for the management of the Lower Kinabatangan region.

Land Tenure   At present the vast majority of land in the region belongs to the oil palm estates, while the local population has become largely disenfranchised. Declining international prices for oil palm products, combined with the problems of flooding and elephant conflicts, have encouraged some operations to sell their estates. As a result of these trends, the highly concentrated land tenure regime today could change in the future.

Economic Incentives   The local threats facing oil palm estates, combined with the growth of ecotourism in the region, provide incipient economic incentives for conservation that are likely to grow in the future.

Background

The catchment of the 560 kilometer–long Kinabatangan River represents approximately one-quarter of the land area of the Malaysian state of Sabah in northern Borneo (fig. 5.7). The Lower Kinabatangan is currently the focus of efforts by WWF to preserve and restore a riparian corridor that provides habitat critical to Asian elephants (*Elaphus maximus*), orangutans (*Pongo pygmaeus*), Sumatran rhinoceros (*Dicerorhinus sumatrensis*), and proboscis monkeys (*Nasalis larvatus*).

The forests of the Lower Kinabatangan contain the largest concentration of orangutans in Sabah, believed to be around 800 individuals. Some 50 mammal species (including 10 primates) and approximately 200 bird species have been recorded in the area. Eight of Malaysia's threatened birds are found in the area, including Storm's stork (*Cicona stormi*) and a number of hornbills. Although there is not particularly high endemism, the area is naturally diverse and maintains species that have been extirpated from many other areas of Sabah. Six basic habitats are represented along the river corridor, including small patches of dipterocarp forest, riparian forest, freshwater swamp forests, limestone forests, ox-bow lakes, and mangroves. The area has 27 ox-bow lakes, more than any other comparable area in Borneo.

Conservation efforts to protect and restore a fragile corridor take place amid a sea of oil palm plantations. These plantations have reduced natural forest cover to a relatively thin (or even nonexistent) strip, and they currently form the matrix of the Lower Kinabatangan river basin. Part of the riparian strip is protected within the 27,000-hectare Kinabatangan Wildlife Sanctuary, although this is highly fragmented and has yet to be established effectively on the ground. Several gaps in the riparian buffer constitute effective barriers to connectivity, including a village and highway that straddle the river at Batu Putih (fig. 5.7). Closing these gaps is critical if the river is to remain an intact corridor for wildlife populations, especially elephants. Invasive vegetation is also a problem in some of the rivers and lakes.

Two issues — flooding and elephant conflicts — are critical to current conservation efforts in the Kinabatangan.

Flooding   Inundation of the Lower Kinabatangan floodplain is an annual event, with at least some areas submerged for between two and thirty-two days. The river's hydrology is strongly influenced by agriculture and logging in the Upper Kinabatangan. The expansion of these activities upstream will continue to alter the flood regime of the river, resulting in higher and more

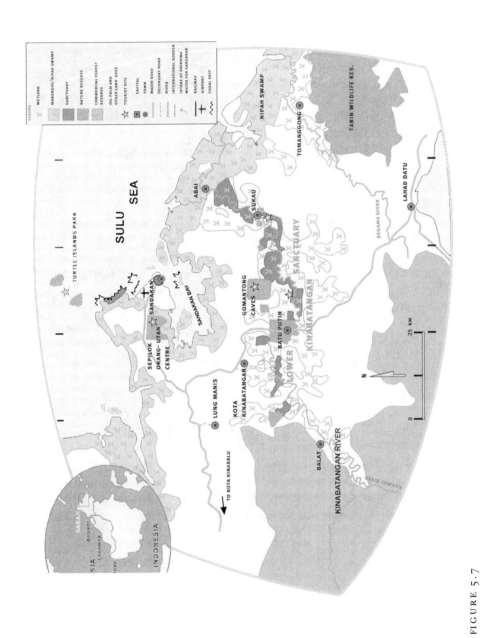

FIGURE 5.7

The Lower Kinabatangan floodplain. The map shows adjacent land use and existing protected areas. Source: Partners for Wetlands.

frequent peak flows. Severe flooding occurred in 1996, and in January–
February 2000 water rose to 14.3 meters above mean sea level, the highest
water level recorded in the region. This flood affected fifteen villages with
more than 3,000 people, and it destroyed approximately 15,000 hectares of
young palm plantations valued at US$11.8 million. Oil palm needs to reach
an age of five years before it can withstand flooding. The apparent increased
frequency of flood events does not bode well for oil palm adjacent to the
river.

Oil palm plantation owners have attempted to build levees. This measure
not only would exacerbate flooding downstream but backfires and holds the
water from the river when exceptionally high floods recede. Furthermore,
levees are expensive to build and are frequently destroyed during intensive
floods. More appropriate strategies for reducing the impact of flooding include (i) maintaining riparian reserves to trap sediment and reduce bank
erosion, (ii) carrying out reduced-impact logging in the upper catchment,
and (iii) allowing logging tracks to recover through natural succession.

Elephant Conflicts   In the Lower Kinabatangan, an estimated eighty elephants migrate from downriver (Sukua) to upriver (Batu Putih) and back in
a cycle that takes about six months. Elephants are attracted to floodplain
habitats where there is plenty of grass. Two factors are contributing to increasing conflict between elephants and people along this route. First, forest
conversion has caused major bottlenecks in the elephant's migration route,
such as at the road at Batu Putih. Second, flooding drives elephants to higher
ground, and this appears to be occurring more frequently as floods intensify.

As a result, elephants are encroaching into oil palm plantations adjacent
to the Kinabatangan River and, in addition to the floods, are causing major
damage to the plantations. Elephants sometimes eat oil palm fruits, pull oil
palm sprouts out and throw them away, and routinely destroy fences. As a
countermeasure plantation workers have attempted to block elephant movements by setting fires. Generally this measure backfires and causes the elephants to wreak greater damage to crops and property (Michael Stüewe,
personal communication).

Possible solutions to elephant conflicts include (i) conserving forest
blocks sufficiently large to maintain the elephant population, and (ii) establishing or maintaining corridors that enable elephants to travel between forest blocks and thereby obtain sufficient food. Both options, and particularly
the latter, probably require fencing or other measures to protect oil palm
plantations.

Fencing can be effective in reducing or eliminating elephant encroach-
ments into estates, but this option is costly. It requires building and maintaining an electric fence with a voltage of 6,000–7,000 V. A trench must be
dug next to the fence and trees must be cleared nearby. These measures
usually provide an effective protection against destruction of fencing by elephants. Nevertheless, floods can wash away the fences and fill up the
trenches, thereby increasing maintenance costs. In short, fencing can be an
effective but usually costly option.

Another option to reduce or eliminate elephant incursions is to manage
elephant population sizes. One approach is sterilization, which is being
tested in Africa. From a conservation standpoint, however, this is equivalent
to killing. Another option is culling. The most common approach in the
Kinabatangan region is translocation, in which the animals are moved to a
local elephant camp at a cost to the oil palm company of US$1,500 per
animal.

Finally, elephants can be trained for use in tourism and other activities.
This option is possible but complex, involving provision of food, water, care-
takers, and veterinary services. Furthermore, elephants in the Kinabatangan
are small and perhaps less interesting to tourists in comparison to larger
elephants used in tourism ventures in Thailand and elsewhere in mainland
Asia.

Initial Activities

Since 1996, WWF Malaysia has been involved in an effort to piece together — through gazetting, effective protection, voluntary changes in land
use, and restoration — a corridor through the Kinabatangan Wildlife Sanctuary. This would eventually connect the upland forests in central Sabah
with mangrove habitats at the coast. Yet there are critical gaps in the chain
of riparian forest that are at risk or in urgent need of restoration.

Ideally, the forested zone along the river would not be less than 500
meters on either side. Malaysian law, however, requires protection of a mere
10 meter–wide buffer zone along rivers — and even this requirement is often
not respected in practice. A much wider buffer zone would be needed to
minimize or prevent elephant conflicts and damage to oil palm plantations.
Much of WWF's efforts have been directed at helping the oil palm industry
in voluntarily complying with the legal requirement for a riparian buffer
and, when possible, surpassing this requirement. Yet there are no official

incentives to conserve or restore natural forests, or even to establish plantations of native forest species, in oil palm estates. As a result, the only incentive
is a negative one: the increasing damage caused by flooding and elephant
incursions.

WWF has begun to support a pilot reforestation effort aimed at testing
approaches to restoring natural forest cover by both natural regeneration and
planting of native species. Initially, the oil palm estate owners were hesitant
to work with WWF, but recent flooding events and increasing elephant
conflicts have convinced some owners that, at least in some areas, oil palm
may not be viable directly adjacent to the river. Following the destructive
2000 flood, three private estates are either allowing forest regeneration or
undertaking tree planting with commercially viable and flood-resistant tree
species on flood-damaged sites. For example, one company (Pontian United
Plantations Berhad) is establishing plantations of native tree species in a 10
meter–wide strip along a 7.5 kilometer stretch adjacent to the banks of the
Lower Kinabatangan (total area planted: 8–10 hectares). This area was badly
damaged by flooding and is considered poorly suited for oil palm. In addition, Borneo EcoTours, a company that offers nature tours and has a lodge
in Sukau, acquired a 25-hectare area at Tenegang Kecil along the banks of
the Kinabatangan. A ceremony initiating tree planting of this area occurred
on June 5, 2000 (World Environment Day), with local and national government officials present.

If successful, these pilot efforts could be expanded into commercially
viable plantations of flood-tolerant timber species. In addition to direct economic returns, the plantations could provide an important buffer against
flooding — and potentially against elephant incursions.

Design

Just as conservation efforts at any site are limited without a full understanding of the larger landscape context, strategies for conserving larger landscapes (such as the Lower Kinabatangan) may be inadequate without an
understanding of the larger ecoregion context. Intermediate-scale planning
efforts should be based on a larger-scale ecoregional vision process that identifies priority landscapes of outstanding biological importance.

No such visioning process has yet occurred for Borneo or any of its constituent ecoregions, including the Northeast Borneo Rainforest ecoregion,
in which the corridor lies. Nevertheless, there have been two conservation

prioritization efforts made for Sabah. The 1981 *Faunal Survey of Sabah* was initiated during a time when forest was relatively widespread in the state. The *Survey* highlighted five priority conservation areas: the Danum Valley, Tabin, Kumambu, Tanjung Linsang, and the Maliau Basin. The Kinabatangan was not among these priorities, but this was before concepts such as area-sensitivity, viable populations, and connectivity were widely recognized, and before most pressures from agriculture had arisen. The 1992 *Sabah Conservation Strategy* was prepared by WWF Malaysia for the then–Ministry of Tourism and Environmental Development, under a package agreement between the Sabah state government and the World Bank. Adopted as official government policy in 1995, the strategy specified the Lower Kinabatangan as an area to be reserved for wildlife. Today it is clear that the Kinabatangan River corridor represents not only a threatened ecosystem in its own right, but a potentially critical connection to upland forest areas in Central Borneo.

There are two primary reasons a large-scale vision should be developed as a basis for work at the landscape level:

- It is necessary to identify where the highest-priority areas actually are in the ecoregion, so that resources are expended strategically and not on low-priority areas.
- A larger-scale, ecoregion perspective allows appropriate (although not precise) delineation of the corridor boundaries.

Regarding the first issue, evaluating the trade-off required in directing scarce conservation resources to one area rather than another requires consideration of four primary factors: biological importance, current conservation status (and how much effort will be needed to achieve conservation objectives), future threat scenarios, and opportunities. The Lower Kinabatangan appears to be a high priority for the Northeast Borneo Rainforest ecoregion, especially when threats and timeliness are considered. The long-term viability of the habitats and species within the Tabin Wildlife Reserve to the southeast and the forest management units upstream (fig. 5.7) may be dependent on establishing and maintaining a connection with the Lower Kinabatangan. Such extensive connectivity will be necessary for the long-term viability of elephants and Sumatran rhinos in the state. The current status of the riparian habitat along the river is good to fair: numerous key species are surviving in the area, but there are some key gaps and the long-term viability of some species is in question. Without active efforts today, however, the future-threat scenario is bleak because pressure from oil palm producers on the thin and fragile corridor is intense. Additional threats are focused on the forest management units upstream, where the owners of the forest concessions are seeking authorization to clear the forest and establish oil palm plantations. Finally, WWF's Partners for Wetlands and AREAS programs are providing critical financial and technical support.

The second issue, involving the limits of the corridor, is more problematic and introduces a series of questions:

- Is the Kinabatangan corridor, currently limited to the Kinabatangan Wildlife Reserve, sufficient to conserve critical habitats, threatened wildlife, and key ecological processes?
- Should the corridor ultimately include the forest reserves upstream and the Tabin Wildlife Reserve to the southeast, with their constituent Sumatran rhinoceros populations (currently believed to be approximately 20 or more animals but with a carrying capacity of approximately 120)?
- How important are the connections between the mangrove forests at the mouth of the Kinabatangan and the upland forests at its headwaters in Central Borneo, and are such connections feasible?

Recently, WWF has developed a more detailed corridor vision, in which the Lower Kinabatangan is part of a much larger landscape corridor connecting the Lower Kinabatangan and adjacent Tabin regions with the Sebuku Sembakung region in Kalimantan. The precise boundaries of this corridor and the criteria used to define it have yet to be determined, and it should be part of the overall, long-term vision of the entire Northeast Borneo Rainforest ecoregion.

Stakeholder Engagement

There are four major stakeholder groups in the region, described below.

Oil Palm Industry   Oil palm has been present in the area since the 1970s, but the swiftest expansion of estates occurred during the period 1985–1990. Plantations now cover 635 sqaure kilometers, or roughly a quarter of the entire Kinabatangan floodplain. The rapid forest conversion process destroyed extensive areas of important wildlife habitat, and it also has contributed to intensified flooding and elephant conflicts.

Local River People   Today the local river people (*Orang Sungai*) are concentrated in fifteen main settlements along the lower reaches of the river. For hundreds of years these people have depended on the river ecosystem for fish and prawns, and on adjacent riverine and upland forests for rattan, beeswax, camphor, edible swiftlet nests, and logs for timber industries that were active in the region from the 1960s to the 1990s. With the spread of oil palm plantations and other forms of agriculture, about 85 percent of the floodplain has been cleared, and the river people have limited access rights to the forest remnants that remain.

Job opportunities also have declined in recent years with the demise of the logging industry — formerly the major source of employment in the region. Some people now are employed on the plantations, although most local employees are in administrative or supervisory posts and most of the labor force is comprised of Indonesian immigrants. Local riverine people potentially could suffer from soil or chemical runoff from the plantations into the river. They also conflict with elephants, as the latter — thwarted in their natural migration along a sometimes broken riverine corridor — trample crops and cause mayhem.

Ecotourism Industry   From 1996 to 2000, tourism increased steadily from roughly 5,000 to 14,000 visitors per year. Most of these concentrate in and around the towns of Abai, Sukau, and Batu Putih, to see wildlife in its natural habitat. There are currently five tourist lodges on the river that provide lodging and wildlife cruises. It is clearly in their interest to maintain and enhance the habitat quality along the river. The five operating tourist lodges provide a source of direct employment for fifty to sixty young people, and indirect employment for an additional twenty to thirty — out of a total estimated population of 4,600 people in the Lower Kinabatangan (according to the 1996 census). Recent trends indicate that tourism is a significant and growing economic force in the region, and soon it may have substantial political clout.

Government   In November 1999, the chief minister of Sabah declared the Lower Kinabatangan Malaysia's first Gift to the Earth. Despite this formal commitment, however, progress in the effective establishment of protected areas in the region has been slow. For example, although the Kinabatangan Wildlife Sanctuary was officially declared in 1996, little has been done to enforce this status. State government agencies are short-staffed and provide low salaries, which impedes effective enforcement of environmental or land-use policies.

WWF Malaysia has deemed it premature to convene all these stakeholder groups to discuss and develop a corridor vision for the Kinabatangan. This is because most of these groups require greater exposure to localized conservation initiatives before they would understand and be supportive of a larger-scale undertaking involving the Kinabatangan corridor.

As described above, however, WWF has helped involve estate owners and tourist companies, in consultation with local governmental agencies, in small-scale reforestation efforts. Likewise, WWF has worked with local communities for many years and currently maintains three community liaison officers who work with tourism and community relations. Finally, in a WWF-sponsored forum in April 2001, representatives from the government, business, and agricultural sectors reached broad agreement about the need for coordinated land-use planning in the region with participation of relevant stakeholders. As a follow-up to this forum, WWF has formulated a conservation and development vision for the region that incorporates biodiversity protection, sound management of natural resources, and strengthening of low-impact economic sectors such as ecotourism and agroforestry (Partners for Wetlands 2001).

Land Tenure

With the rapid expansion of oil palm plantations beginning in the 1980s, the riverine population has lost control over and access to much of its traditional resource base. To date this has not led to significant conflicts with oil palm estates. But future conflicts would have unforeseeable implications for the long-term viability of the Kinabatangan corridor. Improving the livelihood of the disenfranchised riverine people is an important objective in moving the region to a more sustainable footing.

Economic Incentives

The recent flooding episodes in the Kinabatangan Basin — coupled with low prices for oil palm products on international markets and scarce labor availability — have caused some estate owners to leave part of their areas idle, and a few have initiated pilot plantations of flood-tolerant timber species.

If current conditions continue, estate owners may be compelled as a last resort to sell their properties. Several plantations are already up for sale because of the above reasons. One estate, called Abadi Meway, is currently for sale at approximately US$50 million for 5,050 hectares of planted oil palm (five- to ten-year-old trees) and a palm oil factory. If this sale marks the beginning of a trend, land values could drop quickly in the region, thereby providing an important opportunity for securing areas for conservation. This situation provides a potentially strategic conservation opportunity. WWF Malaysia has mapped critical areas for conservation that are currently planted for oil palm and is planning to secure as much land as possible to link the forest corridor on both sides of the river, from the mangrove swamps to upland forest.

As mentioned above, tourism is becoming an increasingly significant economic activity in the region. The 2000 forecast predicted total gross revenues from tourism in the region at US$1.4 million. Tourism already provides employment for a small but growing part of the local workforce. Likewise, revenues from tourism have grown steadily in recent years and are expected to continue doing so in the future, especially because of the aggressive marketing efforts of the Sabah Board of Tourism. These trends indicate that tourism is already an important incentive for conservation and is likely to become more important in the future.

At present there are no official policies providing economic incentives for environmentally sound forms of resource use in the region. Ironically, however, current threats faced by the oil palm industry — in the form of flooding and elephant conflicts — are providing an increasingly strong incentive for estate owners to explore land uses such as plantations of forest species or natural forest regeneration as alternatives to monocultures of oil palm. As mentioned above, these threats — along with the probably far more significant price decreases of oil palm products in international markets — are beginning to lead estate owners to sell their plantations, thereby providing an important opportunity for conservation.

Governance

At present there has been no definition of the kinds of institutional arrangements needed to establish and maintain a corridor along the Kinabatangan. Some relevant questions that would need to be addressed in defining such arrangements are as follows:

- What are the relevant governmental agencies, and what are their roles and responsibilities?

- What are the roles and responsibilities of other actors (e.g., association of oil palm producers or industries, NGOs, association of tourist industries, village organizations, etc.) in the corridor?
- Are there important international donors, banks, etc., that could assist in supporting the corridor?
- What are the obstacles that would have to be overcome for these actors to work in a coordinated fashion (i.e., co-management)?

Conclusions

The Kinabatangan corridor is still in a highly preliminary stage and it will require considerable work to develop a coherent design and implementation strategy. This case involves a number of key issues.

- There is need for a larger ecoregional vision to provide a context for designing the corridor, and the scale of the corridor remains uncertain.
- Stakeholder engagement around a corridor vision is still in a preliminary stage.
- Land tenure issues are shaped by the dominance of oil palm plantations, but this situation could change.
- Potential economic incentives for conservation are provided by current threats to oil palm and increasing tourism in the region.
- The governance arrangements needed for corridor planning, implementation, and maintenance are as yet undefined.

Case 7: The Terai Arc Landscape of India and Nepal

Summary

The Terai Arc landscape spans a 1,500 kilometer length of the India-Nepal border, in a biodiverse yet highly threatened region covered in lowland forests and grassy plains.[25] The landscape-level corridor project was designed to support the movement and survival of several endangered megafauna, including tigers, Asian elephants, and greater one-horned rhinoceroses, as well as to protect watersheds, support ecotourism, and provide harvestable commodities that benefit local communities.

Because the Terai Arc region spans the borders of two nations, it provides particularly good insight into some of the difficulties associated with transboundary management for conservation — a key issue for large-scale, landscape corridors. Located in one of the most economically poor regions of the world, the Terai Arc case also highlights a variety of economic incentives for conservation.

Corridor Design   Several linear corridors across the Terai Arc landscape will link eleven protected areas and national parks and encompass a total area of 34,000 square kilometers, using buffer zones and multiple land-use areas that are compatible with animal dispersal and migration. Implementation will focus on community-based restoration of several degraded "bottlenecks" to animal movement, involving the planting of trees and grasses and allowing overgrazed yet fertile lowlands to regenerate. Human-wildlife conflicts in areas of regeneration will be mitigated by natural or artificial barriers, community watchtowers, and, possibly, human resettlement.

Stakeholder Engagement   Implementation of the Terai Arc corridor will require cooperation from local communities, private landowners, and government agencies in both India and Nepal. Both countries utilize a combination of community and government oversight to manage their forests, and some NGOs are working to increase local communities' control over public lands. Women have been identified as key stakeholders in the region, and conservation organizations are working to increase the participation of women in community decision making and resource management (box 5.12).

Land Tenure   Land in the Terai Arc falls into four basic classifications: government forests, which form the core protected areas; buffer zones and community forests, which are managed in joint arrangements between government and local communities; and private property. In Nepal, 75 percent of the residents of Terai Arc own land, but the vast majority of landowners possess plots that are too small to provide all of their needed natural resources. In both countries community management of forests has had beneficial results. However, government reluctance to grant full management rights to locals has resulted in inefficient use of common resources.

Economic Incentives   Communities in Nepal and India have been willing to participate in restoration and conservation efforts when the results are increased productivity and revenue. Several conservation organizations are working to create incentives that shift private and community land use away from monocrop agriculture, livestock grazing, and commercial forestry, and toward more sustainable and economically viable uses. Such incentives include a percentage of revenues generated from protected areas, development of ecotourism projects, access to government benefits such as livestock vaccines, and technical assistance for forest and grassland restoration.

Governance   While governance arrangements for the Terai Arc landscape are still in the formative stage, the goal is for Nepal and India to manage their portions of the corridor separately but using complementary policies. Both countries will need to establish cooperation in areas such as fire management, poaching control, and training for government managers. Inside each country, effective communication within and among government agencies must be fostered to ensure cooperation on potentially conflicting resource-use policies and management practices.

Background

The Terai Arc is a landscape-level network of linear corridors extending 1,500 kilometers along the borders of India and Nepal (fig. 5.8), encompassing one of the most biologically rich areas on the planet. The Terai Arc connects eleven protected areas in the foothills and southern slopes of the Himalayan Mountains, including the Royal Chitwan National Park (932 square kilometers) and the Royal Bardia National Park (968 square kilometers) in Nepal, and the Rajaji National Park (831 square kilometers) in India. It encompasses an area of 34,000 square kilometers and is home to more than 80 mammalian species, 550 species of birds, 47 species of reptiles and amphibians, 126 fish species, and over 2,100 species of flowering plants (Joshi 2001).

The Terai Arc landscape corridor is designed to provide habitat and facilitate movement for several endangered mammals, including tigers (*Panthera tigris*), greater one-horned rhinoceroses (*Rhinoceros unicornis*), and Asian elephants (*Elephas maximus*). Tigers in particular may be unable to maintain viable populations into the future without increased habitat connectivity. Biologists estimate that the Arc currently supports about 180 breeding tigers (WWF 2001c). Tigers require more than twice the forest area currently available to guarantee the short-term viability of their population. Ultimately, the Terai Arc may need to support about 500 breeding tigers to ensure the species' long-term survival, a population size that would require substantial increases in connectivity between and restoration of forested habitats.[26]

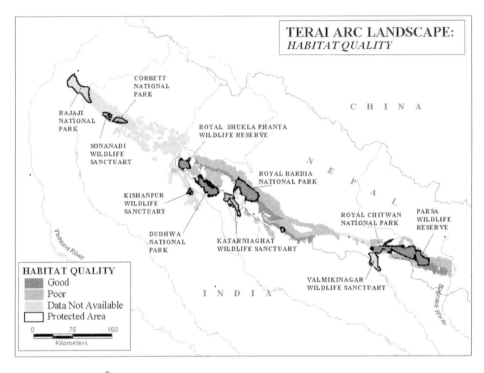

FIGURE 5.8

Habitat quality in the Terai Arc landscape corridor. Source: Adapted after WWF 2001c.

The Terai Arc contains over 75 percent of the remaining forests of the Terai and Churia regions. These forests are crucial for preventing erosion and flash floods, and for recharging the water table in the most productive land in Nepal. Hence, sound management of the Terai Arc is important not only for biodiversity conservation, but also for helping Nepal meet the demands of its growing population for food and forest products.

The human and livestock populations of the Terai Arc pose a direct threat to the diverse wildlife of the area. According to a 1991 census, more than 3.6 million people live in the Terai Arc. This population has grown at ˜3.4 percent annually since 1991, fueled primarily by immigration into the area from neighboring hillside communities (CBS/HMG 1995). The population of the Terai Arc region is expected to reach 6.3 million by 2011, an increase of 75 percent over the 1991 population. About 3.3 million livestock animals also live in the region, with over 75 percent of households completely dependent on agriculture for subsistence. Most people manage their cattle, buffalo, and goats by grazing them in the forests during the day. This practice has reduced or eliminated the regeneration of many forests in the Terai Arc (Chhetri 2001).

The surging population is straining natural resources and increasing habitat fragmentation outside protected areas. Such activities as farming of exotic eucalyptus trees, excessive fires, and overgrazing have substantially degraded many linkage areas. Private landowners increasingly convert forestland for monoculture cash crops, such as hardwood timber (e.g., rosewood and teak) and sugarcane. Poaching for food and for the wildlife trade (particularly for rhinoceros horn and tiger bones used for medicinal purposes in China and Nepal) is also a major problem. As human development eliminates more natural areas, human-wildlife conflicts — particularly tiger attacks and destruction of crops by elephants — have increased. These conflicts lead to revenge killing by people and add to the pressures on animal populations (summarized in box 5.9).

BOX 5.9. MAJOR THREATS TO WILDLIFE IN THE TERAI ARC

- *Accelerating rate of forest degradation and fragmentation.* Between 1978 and 1991, forest cover in the Terai Arc decreased by 12 percent (75,600 hectares).
- *Deforestation and forest degradation.* Increasing forest encroachment and conversion of forests for cultivation by squatters lead to increasing frequency of flash floods and soil erosion.
- *High human population growth rate.* The population in Terai Arc is expected to reach 6.3 million by 2011, largely as a result of immigration of people from the hills adjoining the Terai districts.
- *Grazing pressure.* Large herds of unproductive cattle, buffalo, goats, sheep, and pigs, with a livestock population estimated at 3.3 million animals.
- *Low land holdings per household.* Although more than 75 percent of Terai Arc households are dependent on agriculture for their livelihood, 80 percent of households have less than 2 hectares of land. These households must depend on the forests for their daily supplies of fuel wood.
- *Hunting pressure.* Increased poaching of tigers, elephants, and rhinoceros and other animals, both in protected areas and in forest corridors, because of both hunting for food and the wildlife trade.
- *Human-wildlife conflicts.* Reduction and fragmentation of natural habitats lead to more frequent incursions onto crop lands, increased killing of livestock by tigers, and trampling of crops by elephant herds.
- *Revenge killing.* Increased killing of tigers and elephants because of the problems they generate.

—Adapted from Joshi 2001

The development policies of the Indian and Nepalese governments have left the burgeoning human population of the Terai Arc largely dependent on natural resources for subsistence. Capital investment in infrastructure such as drinking-water treatment facilities and irrigation projects has been made without involving local communities. Government agencies have had insufficient resources to maintain such infrastructure projects, and have not given communities the resources or training to manage them on their own.

In the Nepalese communities of the Terai Arc, local people are willing to manage their own forests but have received little support for their efforts from local or national government agencies. In some villages, the establishment of community forest user groups (CFUGs) has resulted in improvements in forest connectivity and higher local biodiversity. An increasing numbers of communities have applied to national forestry officials for control over their own forests. Yet a slow and cumbersome licensing process that favors wealthy and well-connected landowners has prevented many poor inhabitants from reaping the benefits of community forest management.

While the Nepalese government has designated CFUGs as a legal and praiseworthy concept, officials have been reluctant to put them into practice for fear of losing the revenue streams from these areas (Mingma Sherpa, personal communication). Villagers attempting to enforce antipoaching laws have found their efforts unsupported by local government officials, and many locals have abandoned these efforts. In the words of a villager from the Raptipidit and Gulari Community Forest, "Why should we pay to employ a forest watcher when there is no guarantee that the forest is going to be ours and when smugglers are free to collect all they can?" (Chhetri 2001). As a result, resource extraction from the forests of the Terai Arc continues to pose a major threat to the biodiversity of the region. It will likely continue to do so unless local communities obtain greater control over their local resources.[26]

Design and Implementation

The Terai Arc landscape corridor will be formed by linking existing re-serves using buffer zones and multiple land-use areas that permit animal dispersal and migration. As currently conceived, the corridor does not aim to create new public protected areas. This is because conversion of public land to protected status would remove needed resources from local communities. As discussed above, the human population of the Terai Arc is already large and continues to increase rapidly. Creating new protected areas that further exclude human use of natural resources would only worsen the situation of local people and generate resistance to conservation efforts. Instead, linkage areas will be managed to provide access to subsistence resources, recreation and ecotourism, watershed protection, and other ecosystem services for local communities. Through more sustainable resource use, these areas will also restore and maintain wildlife habitat, movement, and genetic exchange.

Implementation will focus initially on community-based restoration of several degraded forest connections, or "bottlenecks" to animal movement (see fig. 5.8, which shows major gaps in forest cover along the Terai Arc landscape). Priority bottlenecks are identified based on level of degradation (caused mostly by overgrazing of livestock), utility for watershed protection and other ecosystem services, and past wildlife sightings. Linkages along the corridor vary from well-preserved to highly degraded forest habitat. Corridor implementation will prioritize restoration of the most degraded forest bottlenecks. Without this effort, connectivity for wildlife movement along the Terai Arc could be lost within five to ten years (Mingma Sherpa, personal communication). Historical ranges of umbrella species[27] and their prey, recent sightings, animal dispersal models, and satellite images of forest conditions will be used to determine critical linear corridors that need restoration. Research has shown that tigers are capable of crossing gaps in tree cover of up to 5 kilometers in length, so forest restoration need not provide complete coverage. Stepping stones of habitat should be sufficient to support movement by tigers.

BOX 5.10. DESIGNING A CORRIDOR SYSTEM FOR TIGERS
IN THE TERAI ARC REGION

Asia's largest predator, the tiger (*Panthera tigris*), represents a flagship and umbrella species for conservation. Widespread habitat loss and fragmentation have largely confined tigers to protected areas, and only a few populations with tenuous futures exist outside these refuges. Most protected areas are too small to meet tigers' large spatial requirements. As a result, conservation landscapes that link protected areas have been advocated to permit dispersal between core populations, thereby ensuring the long-term persistence of tigers.

continued

The ecological requirements of tigers were used as the basis for a design of a conservation landscape in the Terai Arc along the Himalayan foothills in Nepal and India. Potential habitat linkages were modeled using remote sensing and GIS analysis. The model incorporated broad habitat types, habitat patch sizes, and distances to determine the likelihood of corridor use by tigers. The assumption is that tigers using these pathways will have a greater probability of dispersing successfully from their natal areas. This is because of the better habitat conditions and less travel time spent in suboptimal habitat, relative to alternate routes that present less favorable habitat conditions and more travel time. In the accompanying figure, dispersal potential decreases sequentially from areas in dark gray (level 1 corridors) to areas in gray (level 2 corridors) to areas in light gray (corridor buffer). White areas exhibit little or no dispersal potential.

The corridor system identifies where habitat linkages should be demarcated, and provides recommendations for land use and habitat management to maintain and restore those linkages. The analysis also shows how the ecology and management of wide-ranging species can influence the design of large conservation landscapes.

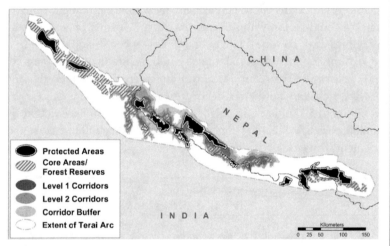

Protected Areas
Core Areas/
Forest Reserves
Level 1 Corridors
Level 2 Corridors
Corridor Buffer
Extent of Terai Arc

— Source: Eric Wikramanayake, Meghan McKnight, and Eric Dinerstein of the World Wildlife Fund, and Anup Josh of the University of Minnesota (USA)

In most cases, restoring priority bottlenecks will entail planting trees in overgrazed areas and/or simply allowing overgrazed but still fertile lowland forests to regenerate. Barriers such as impassable vegetation buffers, rhinoceros trenches, and, in some cases, barbed wire fencing will assist in reducing human-wildlife conflicts. Community watchtowers are also being established to help people ward off crop destruction by elephants. Wherever tigers and humans coexist in close quarters, tiger attacks may continue to be a problem (box 5.11).

BOX 5.11. RELOCATING PEOPLE TO MANAGE ANIMAL MOVEMENT

In some parts of Chitwan National Park, the Nepalese government has relocated entire villages to protect wildlife and avoid human-tiger conflicts. Whether resettlement is a viable conservation option depends on the needs and interests of the local population as well as wildlife-human conflicts. In poor, isolated mountain villages such as Shukla Phanta — where tiger attacks on humans are common, elephants and rhinoceroses damage crops almost daily, and crop yields on the nutrient-poor soils are low — the local community is eager to move. In lowland areas where land is more productive, immigration is heavy and tiger attacks less common, and resettlement holds less appeal for local people. Past government resettlement programs have brought people into Terai from other mountain areas, with negative results for both people and local wildlife.

The challenge when resettling populations to benefit wildlife is to do it in a socially acceptable way. Although it cautions against resettlement, the World Bank has defined a comprehensive Involuntary Resettlement Operational Policy[a] that provides guidelines to be followed when resettlement is deemed necessary.

[a] Available as of 27 August 2004 at: http://wbln0018.worldbank.org/Institutional/Manuals/ OpManual.nsf/whatnewvirt/CA2D01A4D1BDF58085256B19008197F6?OpenDocument.

Other implementation objectives for the Terai Arc landscape corridor include

- promoting forest restoration and management,
- reducing the impact of livestock in forests through increased stall-feeding of livestock,
- promoting alternatives to fuelwood and more efficient use of fuel,
- supporting the establishment of community antipoaching units,
- restoring key species into protected areas, and
- encouraging local participation in reforestation by establishing community forest user groups (as discussed above).

Stakeholder Engagement

Corridor implementation in the Terai Arc will require cooperation from local communities, private landowners, and government agencies in India and Nepal. International conservation organizations such as the World Wildlife Fund (WWF), in partnership with local organizations,[28] have held several meetings convening stakeholders to begin the planning process and to identify government programs, donors, and diplomatic activities that could play a role in Terai Arc conservation. Women are a major focus of stakeholder engagement at the community level (box 5.12).

BOX 5.12. WOMEN AS KEY STAKEHOLDERS

Women play key yet frequently invisible roles in resource conservation worldwide, and their participation and empowerment often enhance the likelihood that conservation initiatives will succeed. As part of a long-term conservation strategy, WWF and partners are working to increase the participation of women in community decision making and resource management in the Terai Arc. Nepalese NGOs such as the King Mahendra Trust for Nature Conservation and Women and Environment have worked with women for more than ten years in some areas. In some regions, village development committees are also involved.

Providing childcare is one of the most important steps for enabling women to become involved in community decision making. Although it is a foreign concept for local people, many women welcome the opportunity to use daycare services, giving them more time to participate in community meetings and income-generating activities. NGOs hire local women as field staff and have used various media, including public theater, to attract women to meetings. Some groups are also providing microcredit loans to women to start businesses, and scholarships to send girls to school.[a] Other groups provide evening literacy classes and training in trades such as tailoring, basketry, poultry farming, and buffalo dairy production.

Greater earning power and participation in decision making also give women more power in their households. In some cultural groups, men have been uncomfortable with these developments, but women have assuaged their husbands' fears by inviting them to attend meetings.

[a] Examples of such groups include the Women's Empowerment Program, USAID/Nepal, Japan-ESCAP, Save the Children, and the government-run Microcredit Project for Women.

Local government land managers are often the most difficult stakeholders to work with in the Terai Arc. Government officers tend to understand conservation only within the context of legal protected areas and other areas under government management. WWF and partners are working to change this perception and increase community control over public lands outside existing protected areas. So far, meetings and study tours are being used to educate government officials about the value of multiple-use, community-controlled conservation areas. WWF and partners are also organizing meetings and study tours to enable community groups to shape corridor development and learn about multiple-use management.

Land Tenure

Land in the Terai Arc falls into four general categories:

- community forests (managed by community forest user groups in Nepal and by government-led joint forest management groups in India);
- buffer zones (managed by buffer zone management councils in Nepal and joint forest management groups in India);
- government forests; and
- private property.

Government-run protected areas that focus mainly on wildlife protection, with limited extractive uses, form the core habitat preserves in the Terai Arc. Government forests support multiple uses in both India and Nepal, including forestry and tourism, and also provide some revenue to local communities.

In India the government co-manages community forests with local people to reduce pressure on natural resources. Local communities are granted a long-term lease on parcels of land that they use for timber, thatch, and other subsistence resources. The local communities and government manage the resource and share the cost equally, while locals are permitted to harvest resources following sound management practices. Similar policies that provide tenure and profit-sharing incentives have been adopted to improve conservation in buffer zone areas. Such community-based management regimes have helped address the subsistence and income-generating needs of local communities. They also provide them with the increased tenure security required to protect the forest resource base. An estimated 36,075 committees in twenty-two states manage 10.24 million hectares of Indian forests under the Joint Forest Management program (DAINET 2000).

In Nepal local people retain greater control over community forests. The Nepalese government grants long-term leases to CFUGs comprised of self-selected community members. The CFUG devises a management plan and, once the plan is approved by the government, is solely responsible for managing the land and owns all resources and revenue generated from it (Joshi 2001).

Conservation and community development NGOs are working to increase community control of public lands in the Terai Arc region. The hope is that secure land tenure will help local communities attain tangible benefits from protected areas and buffer zones. In addition, CFUGs with secure tenure to local resources will resist participation by newcomers and thereby discourage immigration — one of the root causes of human-wildlife conflicts. This could help counter the effects of development assistance and other economic benefits derived from improved land stewardship in areas such as the lowland Terai Belt, which potentially could increase immigration. Preliminary observations suggest that improved tenure might indeed be working as a barrier to immigration (Mingma Sherpa, personal communication).

Much of the land surrounding protected areas has also been privatized, sometimes through squatting. Private control of the land surrounding protected areas can reduce opportunities for new settlers, but will aid wildlife only if landowners can be persuaded to preserve habitat values. Currently, private landowners in many areas are increasingly converting forestland for monoculture cash crops such as high-value timber. As discussed below, local NGOs are working to develop concrete economic incentives to encourage conservation on private land.

While strengthening land tenure for local people in the Terai Arc has the potential to aid development and conservation, it may also create negative social impacts by marginalizing the poorest people in the region. These people lack the resources and education to participate in decision making and do not benefit from profitable CFUGs. If the poor are unable to obtain access to subsistence resources on CFUG and private lands, they may be compelled to meet their basic needs by stealing or move into new areas where their activities may place further pressure on wildlife. Some NGOs and government agencies have initiated programs to help alleviate poverty for marginalized residents of the Terai Arc, including adult literacy classes, microcredit lending, and technical assistance for startup businesses (One Country 2001). Nevertheless, the socioeconomic impacts of land-tenure changes designed to promote conservation may have negative social and environmental impacts. This issue requires further monitoring and analysis.

Economic Incentives

Communities in Nepal have been willing to participate in restoration and conservation efforts when they understand that improved land stewardship can bring new economic opportunities, such as jobs in ecotourism, and increase the productivity and revenue potential of community forestry land. WWF and partners are working to develop concrete economic incentives to shift private land use away from monoculture cropping, and to prevent community forests from becoming conventional commercial forests. Economic incentives being used in the Terai Arc include the following.

Revenue from Protected Areas   Local communities in Nepal receive 30–50 percent of the revenue generated from national parks. This policy has provided substantial revenue for Terai Arc communities. In 2000, tourism to the corridor's two most popular national parks, Royal Chitwan and Royal Bardia, generated about US$1 million. The government of Nepal distributed 50 percent of this revenue to local buffer zone councils, which then determined how to spend the funds for local development (Mingma Sherpa, personal communication).

Ecotourism   Since 1986, communities in the Annapurna region of Nepal have been generating revenues by charging user fees to trekkers and other recreationists. This is done with the help of the Annapurna Project, a non-profit venture led by the King Mahendra Trust for Conservation of Nature in concert with WWF and other partners. Through this program, trekkers pay entry fees when visiting Nepal's national parks and conservation areas. Fees are used for various local conservation and development projects, such as the creation of tree nurseries, establishment of safe drinking water supplies, and the provision of low cost and/or subsidized loans to provide for small-scale hydroelectric generators and solar-powered water heating. Similar revenue-sharing arrangements have benefited communities in the regions surrounding Royal Bardia and Royal Chitwan National Parks in the Terai Arc. With over 70,000 trekkers visiting the Terai Arc each year, some communities have earned considerable revenue in this way. Money raised goes directly to community-based conservation area management committees.

Technical Assistance for Restoration   When allowed to regenerate naturally, lowland areas in the Terai Arc can produce very high yields of grass to be used for thatch, paper, and animal fodder. Grass harvest and appropriate burning of low-elevation grasslands enhance habitat for deer, the primary food source for tigers. CFUGs may develop community forestry programs, which can establish tree plantations on community lands as an alternative to monoculture cash cropping. These plantations produce revenue from the sale of harvested lumber, money that goes directly to the communities. The reforestation also attracts animals such as rhinoceros, which in turn attracts tourism to the area. In partnership with the King Mahendra Trust for Nature Conservation, WWF is providing technical assistance and small loans and grants to support ecologically responsible tourism development such as tour guiding and elephant tours.

Access to Social Services   In Nepal, NGOs are providing field staff to help local communities obtain government benefits and services such as livestock vaccines and family planning materials. Although many such services are already available, lack of resources, inadequate education, and cultural taboos prevent many people in rural areas from accessing them.

Governance

International Governance   Governance arrangements for the Terai Arc are still in the formative stage. The goal is for Nepal and India to manage the region separately but use complementary policies grounded in an integrated conservation and development approach. The transfrontier conservation areas of southern Africa provide a model for this type of international cooperation in conservation and resource management. Botswana, Mozambique, Namibia, South Africa, and Zimbabwe have created several cross-boundary conservation areas. In these areas, neighboring countries have agreed to manage their own resources in ways that are compatible with the goal of long-term sustainability of shared ecosystems.

One of the first priorities for corridor implementation in the Terai Arc is to develop a framework for transboundary governance between India and Nepal. The neighboring countries must establish cooperation in many key areas, including fire management, control of wildlife trade, and poaching. At this time, WWF is supporting transboundary governance by providing aid to control poaching, technical assistance for wildlife management (such as rhinoceros translocation), and training for government land managers and community forest leaders.

National Governance   As discussed above, India and Nepal utilize a variety of management mechanisms for their public and community lands. In both nations, government agencies control and manage public lands while sharing management of community forests with local community groups. Representatives from government agencies tend to resist the creation of multiple-use community forests, and instead favor new, government-controlled protected areas. This has made the use of traditional, top-down approaches to corridor implementation a major challenge in the Terai Arc.

Where community forests have been created, there are some differences in the level of control that Indian and Nepalese communities actually have over their local resources. There is, therefore, variation in the extent to which community forests can be used to create linkages in the corridor. In India, corridors will be established through a combination of joint forest management areas, compatible agricultural uses, and government forests and wildlife reserves, and will be subject to more direct government management than those in Nepal.

Local resource managers in Nepal appear to be better prepared to initiate community-based conservation and development activities than their counterparts in India. In Nepal, the community forest user group program provides a strong policy framework for community control of resources, and there is an ample base of national NGOs with expertise in community development. The Department of National Parks and Wildlife will administer public lands, while community-based organizations will manage community forests. Buffer zone councils consisting of community-elected leaders, aided by local game wardens, will administer buffer areas. India is making progress at this level, and a recent report on tiger conservation makes specific recommendations for local participation in conservation efforts (Johnsingh et al. 2004).

Because the Terai Arc comprises lands with such a diversity of governance and uses, fostering communication between and among government agencies is an important goal of corridor management. Without facilitation by outside groups, agency officials in different departments such as forestry and wildlife often fail to cooperate. Such cooperation is essential for corridor management to be effective in both India and Nepal.

Conclusions

Using an ecoregion approach, WWF and partners have assessed the biological and socioeconomic characteristics of the Terai Arc and identified critical social challenges and opportunities for large-scale conservation. As part of this process, several implementation objectives in the Terai Arc corridor over the next five years include

- participatory delineation of linear corridors in key bottleneck areas;
- increased turnover of community forestry lands from government to local control;
- support of local conservation and development initiatives based on locally defined needs and community management;
- development of field capacity for nature-based tourism, including assessment of probable wildlife viewing routes;
- increased participation of women as key stakeholders in income-generating activities and resource-management decision making at all levels;
- creation of antipoaching units, composed of local people and government officials, to patrol corridors;
- diminished illegal wildlife trade throughout the region; and
- delivery of a signed transboundary agreement on the Terai Arc between India and Nepal.

Case 8: Restoring Landscape Linkages in the Veluwe Region, the Netherlands

Summary

An ambitious program to restore habitats and increase linkages of habitats and landscapes has been under way in the 100,000-hectare Veluwe region of the Netherlands since the mid-1980s.[29] It now serves as a model for (and link to) similar efforts elsewhere in Europe. The following issues are of special relevance in this case.

Corridor Design and Management   Efforts to increase connectivity are under way at various levels. The program has constructed so-called ecoducts, or bridges, to facilitate movement of large mammals like red deer and wild boar over major highways, and a total of fifteen should be in place by 2010. The program is removing selected military facilities, relocating camp sites, and establishing trail networks. Forest management aims to achieve a wide range of objectives — including not only wood production but (primarily) maintenance or restoration of biodiversity and ecosystem services. Habitat restoration ranges from natural regeneration of forests (at zero cost) to transformation of former farmland into valuable wetlands and other habitats by removing nitrogen- and phosphorous-contaminated topsoils (at a cost of US$5,000 or more per hectare).

The program is designing and implementing corridors to link habitats and landscapes within the region, to connect with riverine systems adjacent to the region, and eventually to link with corridor networks elsewhere in the Netherlands and in adjacent countries. A key lesson derived from these activities is that the causes of habitat fragmentation should be addressed before corridor initiatives can be expected to succeed.

Stakeholder Engagement   A collaborative relationship between two catalytic actors in the national and provincial government transformed the Veluwe program, encouraging participation by a wide range of institutions and enabling the definition of a more ambitious set of program goals. This transformation also involved establishing more formal partnerships between the participating institutions, including a governing commission with revolving representatives from key sectors. Public support for the program is increasing as a result of the variety of projects — which involve corridor expansion, outdoor recreation, water management, agriculture, forestry, traffic control, and industry and city planning.

Governance   The collaborative relationship described above was critical to overcome conflicting roles within the national government and between the national and provincial governments. This was accomplished through (i) a series of meetings that provided a voice for all institutional actors and led to a broad statement of goals signed by all participants, and (ii) inclusion of major institutions in a functional governing body for the program.

Background

Efforts to restore landscapes are under way in diverse high-income countries. For highly fragmented landscapes the main strategy is to restore habitat linkages, and at this scale both habitat and landscape corridors can play important roles in different ways. The Netherlands provides a good illustration of the methods of restoring landscape linkages because such efforts have been under way there for some time. Within Holland a notable example is the Veluwe region, where one of today's most carefully planned integrated landscape conservation efforts began in the mid-1980s and now serves as a model for (and link to) similar efforts elsewhere in Europe.

The Veluwe is a hilly area of about 100,000 hectares in central Holland, bounded to the south by a branch of the Rhine and to the east by the Ijssel River. Compared to other regions of the Netherlands, the Veluwe is generally unsuitable for intensive agriculture because of its predominantly dry, sandy, and nutrient-poor soils, which are the result of glaciation.

Human presence in the Veluwe dates from 7,000 years before the present, when forests comprised of beech, birch, and oak covered most of the region. By the year A.D. 1000, overgrazing and agriculture had left the region denuded and subject to intensive wind erosion. In the late 1800s the National Forest Service began reforestation with Scots pine (*Pinus sylvestris*), a native species. Today only about 5 percent of the Veluwe consists of agricultural land, with most of the rest covered by mixed forest (70 percent of the region) and heathland (20 percent of the region and the largest in Europe). The population density of the region is low compared to that of the Netherlands (averaging 450 people per square kilometer), and large areas of the region are completely uninhabited. The Veluwe also contains the country's largest sector for nature-based tourism.

These conditions present an opportunity for restoring a landscape that had been drastically altered over millennia by human settlement and use but is now on the way to recovery. In 1983, the provincial government of Gelderland embarked on a series of initiatives aimed at restoring the landscape. In the beginning these initiatives focused on managing existing forests and heathlands. Forest management provisions were largely production-oriented, reflecting a former policy of deriving 20 percent of forest product needs from forests within the Netherlands. The heathlands, as well as some forests, were targeted for restoration, usually through natural regeneration. Deposition of ammonium — generated from intensive animal husbandry and transported through the atmosphere over long distances — triggers acidification and has degraded some heathland soils, resulting in elimination of characteristic plant and animal species. Mechanical removal of the vegetation and the upper 5 centimeters of the organic topsoil has proved to be an effective way of restoring the native heathland ecosystem and generally costs about US$5,000/hectare.[30] In addition, in the mid-1980s two ecoducts were built over a highway to facilitate movement of red deer (*Cervus elaphus*) and wild boar (*Sus scrofa*) between managed forest areas. These were the first ecoducts in Europe and, probably, the world.

During the 1990s these initiatives began to evolve, and by 2000 the provincial and national governments — together with a wide range of local stake-holders in the region — had defined a far more ambitious ten-year program for the Veluwe region. This program aims to establish a natural landscape with maximum connectivity, which will provide substantial benefits for:

- biodiversity conservation;
- environmental services, such as water storage and regulation of local climate;
- economic activities, such as tourism, recreation, and forestry; and
- cultural preservation initiatives, such as archeological sites and museums.

Furthermore, the program aims to restore environmental linkages between the relatively dry Veluwe and the more humid surrounding areas (including key river ecosystems) and will ultimately establish linkages to landscapes under restoration elsewhere in the Netherlands and in Belgium, Germany, Luxembourg, and France.

Design and Implementation

From a conservation standpoint, the program's main objectives are to:

- restore natural ecosystems, reduce stresses on existing ecosystems (especially forests), and manage them for a variety of environmental values, and
- connect habitats and landscapes at various levels.

These objectives are examined in detail below.

Management   In 1990, the Netherlands established the National Nature Policy Plan, one of the goals of which was to expand forest management objectives beyond production to include other values, in particular environmental services and biodiversity protection. Reflecting these new objectives, today selective cutting takes place at five- to ten-year intervals and can remove only 10–20 percent of the forest cover in a single cut. This form of integrated forest management has been found to maintain yields, encourage natural regeneration, provide environmental services (such as improved habitat for desired species), and cost less than previous practices. Furthermore, scientists and local environmental groups proposed to reintroduce major predators such as wolves (*Canis lupus*) and lynx (*Lynx lynx*). This proposal met with considerable resistance by local landowners and has been abandoned for the moment.

Habitat restoration is a key component of the Veluwe program. Several areas are targeted for restoration efforts: farmlands, heathlands, conversion of single-species plantations into diverse natural forests, and transitions to river systems. Whenever possible natural regeneration is preferred for restoration because of its low cost, even if this requires more time. As mentioned above, however, where soils have been highly contaminated removal of the topsoil is necessary to reestablish proper soil and groundwater quality. Likewise, planting of native species may be necessary to promote forest restoration on sites that have become deficient in native tree composition. Some afforestation takes place for timber production, whenever possible using native species. Costs of habitat restoration vary from zero to US$10,000 or more (in the case of highly degraded farmlands).

Furthermore, the program uses various indirect strategies to restore large blocks of habitat by removing obstacles or reducing their environmental impact. For example, with the end of the cold war numerous military bases in the Veluwe have been closed, and the program is now removing buildings or building complexes from inappropriate locales. One such complex, located in the center of the Veluwe region, encompasses a 60-hectare area that is mostly covered by buildings. In addition, the program is relocating facilities for camping sites and cottages to areas outside of critical habitat blocks, while maintaining or improving a network of trails for hiking, biking, and horseback riding. The program is also developing management plans for farms to encourage restoration of natural areas appropriate for fauna.

Finally, the program is focusing efforts on diminishing the environmental impact of automobiles. One strategy is to discourage the use of secondary roads — which are increasingly being used to avoid heavy traffic on major highways — by reducing speed limits to 60 or even 30 kilometers per hour. Another is to eliminate some secondary roads, although at present there is little local support for this measure. A third strategy is to encourage visitors to walk or use bicycles or horsecarts instead of touring the Veluwe by car. There will be five special parking areas where visitors obtain tour guides and alternative forms of transport, such as rental bicycles or horsecarts. Because they enable people to transfer from their car to environmentally friendly transportation, these parking areas will be called "nature transferia."

An important insight derived from these activities is that the causes of habitat fragmentation must be addressed before corridor efforts — whether at local or international scales — can be expected to succeed.

Connectivity   Efforts to promote connectivity are taking place at various scales. In addition to the three ecoducts built to date, fifteen are planned for the next years (fig. 5.9).[31] These ecoducts span highways and are covered by herbaceous vegetation and fringed by trees or shrubs. Similar to underpasses constructed in the Florida everglades to permit movement of the highly endangered Florida panther (see box 5.2), ecoducts provide an alternative solution for restoring animal movement through human infrastructure. Construction of each ecoduct costs US$4–5 million, a figure that has remained stable since the first one was constructed in the mid-1980s. Long-term monitoring of the first ecoduct by means of a simple sandbox has shown that deer use them for movement and browsing. Wild boars use the ecoducts as well, although these animals are less critical than deer and will sometimes cross a conventional bridge. Many other animals have been found to traverse ecoducts, including foxes, rabbits, martens, rodents, and lizards. Based on experience to date, ecoducts are considered to be an effective way to promote undisturbed movement by a wide range of animals.

At a larger scale, as described above, the program is increasing habitat continuity by removing obstructions such as military facilities and camping sites, and also by reducing the barrier effect of secondary roads. At a regional scale, restoration efforts are focusing on degraded habitats along ecological gradients, beginning with the xeric upland soils of the Veluwe and culminating in river systems surrounding the region — such as the Rhine and Ijssel Rivers. Such targeted efforts will restore a wide variety of ecosystems within a relatively small area. These efforts are also strategic because river ecosystems and their surroundings provide critical movement corridors for animals and plants, and they can help restore key ecological services, such as reducing soil erosion caused by flooding. Habitat restoration in floodplains is transforming former farmlands into marshes. In addition, restoration of areas in and adjacent to river systems opens up new opportunities for recreation and tourism.

Finally, as part of the Netherlands' 2000 Nature Policy Plan, the government has set a nationwide target of placing an additional 200,000 hectares in protected areas by 2020; these will be linked by a network of seven large corridors 500–1,000 meters wide. The corridors planned or under implementation in the Veluwe will be an integral part of this national network, requiring linkages over distances of 30–50 kilometers.

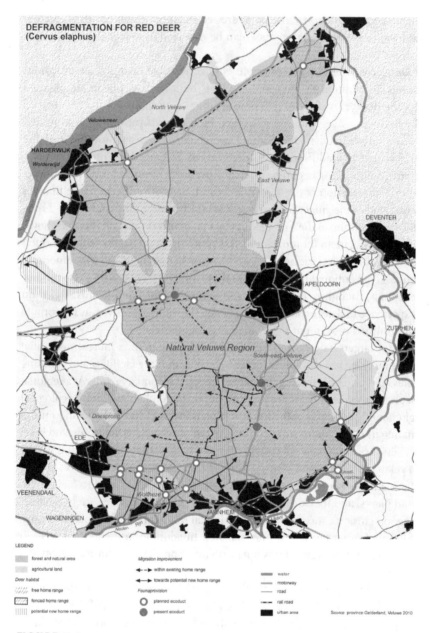

FIGURE 5.9

Current and future locations of ecoducts to facilitate animal movement in the Veluwe region.
Source: Vreugdenhil 2000.

At the international level and in accordance with the European Community's Habitat and Bird Directive, the Netherlands is committed to protecting 10 percent of each of its major habitats. Likewise, within the framework of its nature conservation policy established in 2000 (www.europa.eu.int/comm/environment/nature/home.htm), the European Community is committed to establishing an ecological network to link ongoing corridor initiatives in neighboring member countries. As one of the first such initiatives in Europe, the Veluwe program provides a useful model and case study for corridor efforts elsewhere in Holland and in Europe.

Accomplishments to Date   To date three ecoducts have been constructed, military facilities and exercise areas have been removed, the first recreational cottage area is being relocated, almost all dirt roads are closed for motorized traffic, traffic is diminishing on secondary roads, and new trail networks are being established. Forest management aims to achieve diverse objectives, to be defined by zoning for (i) strict protection of ancient woodlands containing oak trees up to two thousand years old, (ii) natural regeneration without active human interference, and (iii) timber production through limited harvests every five to ten years. Habitat restoration is under way at strategic sites throughout the region, and corridor initiatives are planned or implemented at a wide range of scales as part of a coordinated network. Finally, the program has taken a number of steps to attract broad public interest and support.

Stakeholder Engagement

Establishing a far more ambitious program in the Veluwe has required an ongoing process of consultation among an increasingly diverse array of stakeholders, and today this process takes up most of the program coordinator's time. In its initial phase the program was essentially a provincial initiative that counted on collaboration by local governments and NGOs, with little support from the national government. National government support increased substantially in 1995, however, when a senior technical staff person from the national government began to establish a collaborative relationship with the provincial government's local program coordinator. This single relationship transformed the program, because of the commitment of these two actors and their catalytic role within their respective institutions. They organized a workshop that led to a series of scientific surveys and provided a basis for a long-term regional program presented in 2000 —

Veluwe2010. This program is based on a declaration signed by all involved stakeholders (six ministries, the provincial government, eighteen local communities, several NGOs, and private landowners). The declaration and Veluwe2010 received a great deal of coverage in the local and national press and on radio and television. The declaration proved to be a turning point, as it empowered a wide variety of key institutions (including six national ministries) to define a new program vision and their respective roles in achieving it.[32]

Today the program's institutional arrangements are more formalized than in the past. Broad strategic directions are defined by the Veluwe Commission, which meets every six weeks and is composed of fifteen representatives from governmental agencies, NGOs, and local communities. A smaller and more technically oriented working group meets weekly and is responsible for developing projects to be implemented under the program.

To attract increased public support and participation, these projects touch on practically the entire range of human interests and activities in the region — including corridor establishment and maintenance at various scales, recreation, tourism, environmental education, forestry, agriculture, cultural preservation, city planning, industrial activities and infrastructure, and water management.

Tenure Issues and Zoning

The different forms of land use planned for the Veluwe will be defined based on an ongoing zoning exercise. Most of the region will be designated for environmental protection or forest management, and the program also aims to restore natural habitat in areas currently used for agriculture. In addition to significant expansion of protected areas, the national government requires registration and restricted use of remnants of seminatural habitat on private lands, which are major reservoirs of biodiversity in the Netherlands today. The landowners, who are mostly farmers, receive cash subsidies for the environmentally sound management and protection of these habitats, with the level of subsidy depending on the degree of work involved.

Finally, a key objective of the Veluwe program is to establish a national park over the entire region. The issue of how this park will restrict resource use is of fundamental concern to local landowners. Two distinct concepts are currently being debated and will have important implications for the future of the Veluwe landscape. Under the first concept, equivalent to the

World Conservation Union's (IUCN) category 2,[33] the primary objective of a national park would be biodiversity protection, thereby requiring strict limitations on resource use. The second concept, equivalent to IUCN category 5, would be a multiple-use area where resource-use activities such as agriculture and forestry could continue. Most conservation organizations and a large segment of the region's urban population support the first concept (i.e., category 2), while farmers and other resource users favor the second (category 5). As part of the program's ongoing activities, each of these concepts is being assessed to determine which would be more appropriate for the specific conditions of the Veluwe region.

Economic Incentives

As mentioned above, the national government provides cash subsidies to landowners who conserve biodiversity. The level of subsidy depends on the degree of work involved. For example, forest management requires less work than yearly mowing of moist grasslands and thus receives a lower subsidy.

The anticipated costs of implementing landscape restoration initiatives in the Veluwe through 2010 are estimated at US$300 million. A substantial part of this funding is expected to come from the European Community, the national government, and the provincial government. Additional funding will be provided by local governments and NGOs. Recent data from the National Agency for Statistics, which has helped justify increased support from all levels of government, shows that

- a total of 30 million people visit the region each year,
- these visitors spend about US$1 billion, and
- recreation alone supports 22,000 jobs in the region.

In short, nature-based activities are the most important segment of the regional economy. This fact helped convince decision makers that conservation is important for both environmental and economic reasons.

Attempts to gather funding from the private sector have had little success to date, because most representatives consulted claim that government taxes in the Netherlands are relatively high and should cover the program.

There are few direct economic incentives for conservation by private landowners in the Veluwe. The owner of a property measuring 5,000 hectares in the center of the Veluwe, called National Park de Hoge Veluwe,

however, has established a well-run commercial operation for visitors, including an information center, a museum, and well-maintained trails and
sites for viewing wildlife. In addition, Dutch law enables private landowners
who open their property for public recreation to receive a partial discount
in their land taxes. But even though they provide access for streams of visitors
to local recreational and tourist facilities, landowners receive no income
from those businesses. An umbrella organization for private sector interests
in the Veluwe, which participates in the program's working group, is currently investigating alternative incentives that would encourage and help
defray the expenses of conservation by local landowners.

Governance

The national and international policy contexts for the Veluwe program
are described in the section on corridor design and implementation. Likewise, the institutional arrangements that now govern the program are described under the section on stakeholders. A clearly defined policy context,
as well as clear roles and responsibilities of the main institutional actors,
bodes well for the future of this program.

The national government has established an ambitious set of conservation
targets, both within the Netherlands and in tandem with similar efforts in
other countries. These targets complement and build on those of the Veluwe
program by establishing a nationwide ecological network comprising core
areas and corridors. This is in turn integrated with a Europe-wide network.
In addition, the national government is currently the major source of funding for the program through various ministries. The provincial government
coordinates the program at the local level, plays a critical role in building
both local and national support, and provides additional funding. Both the
national and local governments participate actively in defining the program's
strategic directions. Finally, a wide range of local institutions — including
townships, community organizations, environmental NGOs, and business
interests — participate in defining strategic directions and in preparing and
executing projects.

In the design of the Veluwe Commission, a key factor has been to strike
a balance between representation and decision-making capacity. Consequently, although many more institutions are involved in the program, mem-
bership in the commission is limited to fifteen representatives. Every effort
has been made to represent all major stakeholder groups, and revolving

representation allows a larger group of institutions to play a decision-making role. That role was also a part of the new program vision, when participation of the widest possible range of institutions was encouraged and broad consensus for an ambitious set of objectives was forged.

Conclusions

This case reveals a number of key issues involving the design and implementation of the Veluwe corridor.

- The Veluwe program is designed to promote connectivity through various initiatives and at diverse scales, and it is linked to national and international initiatives with similar aims.
- Stakeholder engagement has been a critical component of the program since its outset.
- A controversial zoning issue involves the conceptual definition of a national park in the region and how this will restrict resource use.
- Economic incentives are provided to farmers who restore natural habitat in areas currently used for agriculture.
- The program provides a noteworthy example of co-management, in which diverse stakeholder groups have well-defined roles and responsibilities and are represented in the decision-making process.

Notes

1. Introduction

1. "Inbreeding depression" is mating among close relatives, which produces few offspring, which in turn tend to be weak or sterile.
2. For example, in the United States other than Alaska, even the largest parks are "small habitat islands" for large carnivores such as cougars (which require 60,000 square kilomenters), bears (6,000 square kilometers), and bobcats (2,000 square kilometers) (Harris and Gallagher 1989).
3. Fragmentation can affect nutrient cycling by exposing fragmented habitats to more intensive wind and water currents or by reducing the number of species, such as beavers or burrowing animals, that contribute to nutrient cycling.
4. An ecoregion is a relatively large unit of land or water that contains a distinct assemblage of natural communities. In comparison to smaller-scale approaches, ecoregion conservation is more suitable for planning biodiversity protection because ecoregions: (i) address the needs of the populations and species that require the largest areas, (ii) represent a comprehensive set of biogeographically related communities and ecosystems, (iii) encompass major driving ecological and evolutionary processes that create and maintain biodiversity, and (iv) permit identification of the best places in which to invest conservation efforts (Dinerstein et al. 2000).
5. Here, "governed" does not refer merely to control by public agencies, but to the broader concept of governance that can involve civil society as a whole. To be sustainable over the long term, corridors require the support and involvement of such diverse interest groups as resource owners and users, businesses, scientists, environmentalists, and the like, in addition to public officials and

agencies. Ideally, each of these groups will have a clear and complementary role in corridor planning and implementation.

2. Conceptual Foundations of Corridors

1. There are three major habitat elements: *habitat patches* or *fragments*, together with *corridors*, are embedded in a *matrix*. (See figure 2.1 for a visual representation of the relationship of these three landscape elements.) The shifting patterns of these elements comprise a landscape *mosaic* (Forman and Godron 1986).
2. See Debinski and Holt (2000) for an analysis of habitat fragmentation experiments.
3. Corridors designed primarily for recreation or aesthetics are often referred to as "greenways" or "greenbelts" (Smith and Hellmund 1993). This book focuses on corridors designed *primarily* for conserving biodiversity or related ecological processes that support biodiversity conservation. This primary focus, however, should not exclude secondary objectives such as recreation or aesthetics — especially if these can help build public support without undermining conservation.
4. The goals of ecoregional conservation include: (i) protection of special elements — including viable populations of key species, habitats, and ecosystems; (ii) representation of species, habitats, and ecosystems; and (iii) conservation of critical ecological processes.
5. The threats posed by disease and parasitism, however, are likely to increase as organisms are crowded into shrinking habitats. Modeling has shown that, under a narrow range of conditions, corridors may dramatically increase the probability of metapopulation extinction through spread of disease (Hess 1994). Because they can increase the ability of individuals to move between patches, corridors might enhance the transmission rate of a disease throughout a metapopulation.

4. Corridor Implementation

1. To put this figure in perspective: Tourism-related activities in the Veluwe generate 1 billion euros in annual revenues. Accordingly, the 10 million euros in planned expenditures for corridor-related initiatives in the region, designed to help restore natural environments and processes such as migration (which are major draws for tourists), are equivalent to 0.1 percent of the gross revenues from tourist-related activities over a ten-year period.
2. Additional ideas for funding corridors in the United States are provided by Little (1990).

5. *Case Studies*

1. Scores of so-called ecological networks worldwide have been described by Bennett and Wit (2001), with a recognized information bias toward corridors located in Europe. These authors use a restricted definition of corridors, equivalent to the definition of linear corridors in this document, and their use of "reserve networks" has similarities with the concept of landscape corridors used here. Consequently, it is reasonable to expect that there are substantially more corridor initiatives (including both linear and landscape corridors, as defined in chapter 2) under way worldwide.

2. Case 1 was prepared in collaboration with Denise Rambaldi, Golden Lion Tamarin Association (Brazil).

3. Species numbers available at http://www.biodiversityhotspots.org.

4. As a typical group of 5–6 individuals requires 45 hectares of forest habitat, an area of 25,000 hectares is considered sufficient to sustain populations totaling 2,000.

5. Some of the native species include: (i) pioneer trees such as "embaúba" (*Cecropia hololeuca*), "crindiúva" (*Trema micrantha*), "jacaré" (*Piptadenia paniculata*), "aroeira" (*Schinus terebinthifolius*), "carrapeta" (*Guarea guidonia*), and "cafezinho-do-mato" (*Caesaria sylvestris*); and (ii) fruit trees such as "jenipapo" (*Genipa americana*), "jaboticaba" (*Myrciaria trunciflora*), "grumixama" (*Eugenia brasiliensis*), and guava, or "goiabeira" (*Psidium guajava*). Unfortunately for the corridor initiative, Brazilian legislation currently prohibits use of native species for timber harvesting, even under a sustainable management plan. Given this limitation, an interesting design for the corridor could include an inner core of native species protected by an outer strip of *Eucalyptus* — which is widely planted in the region as a source of wood.

6. This case was prepared by Sophia Bickford, University of Maryland (USA), and Jennifer Reed, World Wildlife Fund (USA).

7. TNC's Adopt an Acre program provides "critical rainforest acquisition and management funds, enabling TNC and its partners to achieve their mission of protecting biological diversity." The Parks in Peril program operates in threatened national parks and reserves of "global significance, creating on-site logistic and scientific capacity to successfully manage" the ecosystems within them (http://www.nature.org).

8. However, use of plantations of shade-grown coffee or cocoa as sites for wildlife conservation may be possible only over relatively short periods because often farmers transform these sites into open plantations or cattle pastures.

9. This case was prepared with the assistance of Richard Hilsenbeck, The Nature Conservancy, and Sophia Bickford, University of Maryland (USA). Average land cost was approximately US$5,500 per hectare. Other P2000 costs supported scientific analyses and site assessments to prioritize the best areas for

conservation, and due diligence (appraisals, surveys, etc.) to ensure that the land purchases were properly executed.

10. Big Cypress National Preserve, Florida Panther National Wildlife Refuge, Fakahatchee Strand State Preserve, and Picayune Strand State Forest. This case was prepared by Sophia Bickford, University of Maryland (USA).

11. This case was prepared by Sophia Bickford, University of Maryland (USA).

12. In 1995, nonlabor income constituted 36 percent of all income for the U.S. portion of the Y2Y Corridor, twenty times as much as farming and ranching combined and eleven times as much as oil, gas, mining, and forestry combined.

13. As much as 90 percent of the lowland riparian habitat in the Y2Y Corridor has been destroyed by grazing, logging, and other extractive activities.

14. BCEAG members come from the municipal district of Bighorn, the town of Canmore, Banff National Park, and the Alberta provincial government.

15. The MKMA region is heavily dependent on resource-based industries. In 1996 forestry, mining, fishing, and trapping accounted for 48 percent of the economy in the Northern Rockies (Fort Nelson) region and 37 percent in the Peace River (Fort St. John) region (www.bcstats.gov.bc.ca). Permanent employment in oil and gas accounts for about 20 percent of local employment, and the timber sector accounts for 40 percent of employment in the town of Fort St. John (www.luco.gov.bc.ca/lrmp/mk/muskwa.htm).

16. Mining representatives contend that land-use planning should not restrict mining because this activity is highly site specific and takes up a small portion of the land area.

17. This case was prepared by Sophia Bickford, University of Maryland (USA), and Zach Stevenson, World Wildlife Fund (USA).

18. Two other exceptionally rich coniferous ecoregions are the southeastern conifer forests of North America and the forests of the Primorye region of the Russian far east.

19. Fir and cedar forests, ponderosa pine forest, rocky juniper habitat, and Garry oak/mountain mahogany habitat converge in the CSNM (Mitchell 2001).

20. Based in Corvallis, Oregon (www.consbio.org), CBI is a leading center for assessing protected areas and making global assessments of areas worthy of conservation.

21. There are no active mining or oil and gas claims in the CSNM region, and the timber extracted is insignificant to the industry.

22. The CSNM is located in Jackson County.

23. The BLM interim management plan outlaws timber harvest on public land, restricts vehicle access on BLM roads, and has closed roads and trails in some sensitive areas while the management plan is being developed.

24. This case was prepared with the assistance of John Morrison, World Wildlife Fund (USA), and Cede Prudente, World Wildlife Fund (Malaysia).

25. This case was prepared by Sophia Bickford, University of Maryland (USA), and Jennifer Reed, World Wildlife Fund (USA).

26. Assuming that approximately 20 percent of the tiger population consists of breeding females, a population of 900 tigers is likely to exist in the Terai Arc. Current estimates of forested land area in the Terai Arc of 5,460 square kilometers equal less than half of the 12,600 square kilometers that a population of 900 tigers would require for survival. Based on a minimum viable population of 500 (Soulé 1987), and Dinerstein et al.'s conservative estimate (2000) of a breeding unit home-range size of 14 square kilometers, ensuring the long-term genetic viability of a tiger population in the Terai Arc could require a total of 3,000 animals. This in turn could require 42,000 square kilometers of forested habitat, preferably in large blocks.

27. As discussed earlier, umbrella species — such as tigers, elephants, and rhinoceros — support a suite of other species through activities such as population control of herbivores (thereby maintaining grassland habitat) or creation and maintenance of new habitats (such as opening forest clearings).

28. Local partner organizations in the Terai Arc include the Nepal Department of National Parks and Wildlife Conservation, Nepal Department of Forests and Soil Conservation, Wildlife Institute of India, Women in Environment, King Mahendra Trust for Nature Conservation, CARE, and the Environmental Camps for Conservation.

29. This case was prepared in collaboration with Bram Vreugdenhil, Veluwe regional government (the Netherlands).

30. This mechanical removal of organic topsoil imitates an ancient land-use system known as sod-cutting, which was done to reverse the natural succession to forest and establish open habitats.

31. In figure 5.9 the dotted arrows indicate potential opportunities for interchange between areas of currently fragmented but suitable habitat. The solid arrows indicate potential opportunities for interchange between areas of existing suitable habitat and of potentially new suitable habitat.

32. A well-illustrated publication (*Veluwe2010, een Kwaliteitsimpuls!* — *Veluwe-2010: An Infusion of Quality*; Vreugdenhil 2000) describes in detail the program's objectives and its ongoing and planned initiatives. This publication targets a broad audience, including the European Union, the national government, regional and local governments, NGOs, the local business community, and other stakeholders.

33. IUCN categories have been adjusted to account for the highly transformed landscapes characteristic of Europe.

References

Akçakaya, H. R., and M. G. Raphael. 1998. Assessing human impact despite uncertainty: Viability of the northern spotted owl metapopulation in the northwestern USA. *Biodiversity and Conservation* 7:875–894.

Aldous, V. 2001. Details thin on future of monument. *Ashland Daily Tidings.* May 4, 1.

AREAS II Technical Working Group. 2001. *Workshop Report, AREAS Program.* Ho Chi Minh City, Vietnam: World Wildlife Fund.

Ashland Daily Tidings. 2001. Opinion: Monument in the shadow of the giant. *Ashland Daily Tidings.* April 27, 4.

BCEAG (Bow Corridor Ecosystem Advisory Group). 1999. Wildlife corridor and habitat patch guidelines for the Bow Valley. Alberta Environment, Calgary, Canada.

BC Stats. 2001. Indicators of economic hardship for Region 55–Peace River and Region 59–Northern Rockies. http://www.bcstats.gov.bc.ca.

Beier, P. 1992. A checklist for evaluating impacts to wildlife movement corridors. *Wildlife Society Bulletin* 20:434–440.

——. 1993. Determining minimum habitat areas and habitat corridors for cougars. *Conservation Biology* 7:94–108.

——. 1995. Dispersal of juvenile cougars in fragmented habitat. *Journal of Wildlife Management* 59:228–237.

Beier, P., and R. F. Noss. 1998. Do corridors provide connectivity? *Conservation Biology* 12:1241–1252.

Bennett, A. F. 1991. Roads, roadsides, and wildlife conservation: A review. In Saunders and Hobbs 1991b, 99–118.

——. 1999. *Linkages in the Landscape: The Role of Corridors and Connectivity in Wildlife Conservation.* Gland, Switzerland: World Conservation Union (IUCN).

Bennett, G., and P. Wit. 2001. *The Development and Application of Ecological Networks: A Review of Proposals, Plans, and Programmes.* Amsterdam: Advice and Research for Development and Environment (AIDEnvironment) and the World Conservation Union (IUCN).

Bierregaard Jr., R. O., T. E. Lovejoy, V. Kapos, A. A. dos Santos, and R. W. Hutchings. 1992. The biological dynamics of tropical rainforest fragments. *Bioscience* 42:859–866.

Binford, M. W., and M. J. Buchenau. 1993. Riparian greenways and water sources. In Smith and Hellmund 1993, 69–104.

Brandon, K., K. Redford, and S. Sanderson, eds. 1998. *Parks in Peril: People, Politics, and Protected Areas.* Washington, D.C.: Island Press.

Burkey, T. V. 1997. Metapopulation extinction in fragmented landscapes: Using bacteria and protozoa communities as model ecosystems. *American Naturalist* 150:568–591.

CBS/HMG. 1995. Population monograph of Nepal. His Majesty's Government of Nepal, National Planning Commission Secretariat, Centre Bureau of Statistics, Ramshah Path, Kathmandu.

CBTC (Asociación de Organizaciones del Corredor Biológico Talamanca Caribe). 2001. *Planificación para la conservación de sítios PCS: Nombre del sítio Talamanca (sector terrestre).* Costa Rica: The Nature Conservancy.

Chape, S., S. Blyth, L. Fish, P. Fox, and M. Spalding. 2003. *United Nations List of Protected Areas.* Gland, Switzerland: IUCN; Cambridge: UNEP-WCMC.

Chhetri, R. B. 2001. A socio-economic situation analysis of the proposed Terai Arc Landscape Program: Local level perspectives from the three bottleneck areas. Unpublished document.

Chomitz, K. M., E. Brenes, and L. Constantino. 1998. *Financing Environmental Services: The Costa Rican Experience.* Economic Notes 10. Washington, D.C.: World Bank.

Conservation International. 2000. *Designing Sustainable Landscapes: The Brazilian Atlantic Forest.* Washington, D.C.: Center for Applied Biodiversity Science, Conservation International.

———. 2001. Population status, ecology, and transboundary movements of elephants in the Okavango–Upper Zambezi Transfrontier Conservation Area. Unpublished document.

Core Technical and Planning Team. 2001a. Florida peninsula ecoregional plan. The Nature Conservancy (Tallahassee) and the University of Florida GeoPlan Center Gainesville, Florida, USA. Draft plan.

———. 2001b. Tropical Florida ecoregional plan. The Nature Conservancy (Tallahassee) and the University of Florida GeoPlan Center (Gainesville). Draft plan.

Cox, J., R. Kautz, M. MacLaughlin, and T. Gilbert. 1994. Closing the gaps in Florida's wildlife habitat conservation system. Office of Environmental Services, Florida Game and Fresh Water Fish Commission.

DAINET (Development Alternatives Information Network). 2000. Joint forest management: What is it? http://www.jfmindia.org.

Debinski, D. M., and R. D. Holt. 2000. A survey and overview of habitat fragmentation experiments. *Conservation Biology* 14:342–355.

Defazio, P., E. Blumenaur, D. Hooley, and D. Wu. 2001. Letter to Department of the Interior Secretary Gail Norton. May 15. U.S. House of Representatives, Washington, D.C.

DellaSala, D. 2000. WWF comments on the DEIS for the CSEEA. WWF, Ashland, Oregon.

——. 2001. Statement in support of Cascade–Siskiyou National Monument to the Jackson County Natural Resources Committee. Presented at June 12 hearing in White City, Oregon.

Diamond, J. A. 1975. The island dilemma: Lessons of modern biogeographic studies for the design of natural reserves. *Biological Conservation* 7:129–146.

Dietz, J. M., L. A. Dietz, and E. Y. Nagagata. 1994. The effective use of a flagship species for conservation of biodiversity: The example of the lion tamarins in Brazil. In *Creative Conservation: Interactive Management of Wild and Captive Animals*, ed. P. J. S. Olney, G. M. Mace, and A. T. C. Feistner, 32–49. London: Chapman and Hill.

Dinerstein, E., G. Powell, D. Olson, E. Wikramanayake, R. Abell, C. Loucks, E. Underwood, T. Allnutt, W. Wettengel, T. Ricketts, H. Strand, S. O'Connor, and N. Burgess. 2000. A *Workbook for Conducting Biological Assessments and Developing Biodiversity Visions for Ecoregion-Based Conservation*: Part I. Washington, D.C.: Conservation Science Program, World Wildlife Fund.

Edwards, L. 2001. Interview with Jackson County Commissioner Sue Kupillas. Medford, Oregon: Jefferson Public Radio, Jefferson Daily program June 14.

Ehrlich, P. R., and I. Hanski, eds. 2004. *On the Wings of Checkerspots: A Model System for Population Biology*. New York: Oxford University Press.

Fahrig, L., and G. Merriam. 1994. Conservation of fragmented populations. *Conservation Biology* 8:50–59.

Fattig, P. 2001a. Some residents of monument aim to save it. *Medford Mail Tribune.* April 4, 1.

——. 2001b. Feds delay monument plan. *Medford Mail Tribune.* May 18, 1.

Ferraro, P. J., and A. Kiss. 2002. Direct payments to conserve biodiversity. *Science* 298:1718–1719.

——. 2003. Response to: J. A. A. Swart. Will direct payments help biodiversity? *Science* 299:1981–1982.

Ferraro, P. J., and R. D. Simpson. 2002. The cost-effectiveness of conservation payments. *Land Economics* 78:339–353.

Florida Department of Environmental Protection and the Florida Greenways Co-ordinating Council. 1998. Connecting Florida's Communities with Greenways and Trails: The Five-Year Implementation Plan for the Florida Greenways and Trails System. Tallahassee.

Florida Department of Environmental Protection, Division of State Lands, Office of Environmental Services. 2000. A Ten-Year Acquisition Plan — Preservation 2000. Conservation and Recreation Lands 2000 Report.

Forman, R. T. T. 1995. *Land Mosaics: The Ecology of Landscapes and Regions*. Cambridge: Cambridge University Press.

Forman, R. T. T., and M. Godron. 1986. *Landscape Ecology*. New York: Wiley.

Forsman, E., R. Anthony, J. Reid, P. Loschl, S. Sovern, M. Taylor, B. Biswell, A. Ellingson, C. Meslow, G. Miller, K. Swindle, J. Thrailkill, F. Wagner, and E. Seaman. 2002. Natal and breeding dispersal of northern spotted owls. *Wildlife Monographs* 149:1–35.

Foster, M. L., and S. R. Humphrey. 1995. Use of highway underpasses by Florida panthers and other wildlife. *Wildlife Society Bulletin* 23:95–100.

Friend, G. R. 1991. Does corridor width or composition affect movement? In Saunders and Hobbs 1991b, 404–405.

Gadd, B. 1998. The Yellowstone to Yukon landscape. In Harvey 1998, 9–18.

Gailus, J. 2000. *Bringing Conservation Home: Caring for Land, Economies, and Communities in Western Canada*. The Sonoran Institute and the Yellowstone to Yukon Conservation Initiative.

Gascon, C., and T. E. Lovejoy. 1998. Ecological impacts of forest fragmentation in central Amazonia. *Zoology* (Wena, Germany) 101:273–280.

Gascon, C., G. B. Williamson, and G. A. B. da Fonseca. 2000. Receding forest edges and vanishing reserves. *Science* 288:1356–1358.

Gonzalez, A., J. H. Lawton, F. S. Gilbert, T. M. Blackburn, and I. Evans-Freke. 1998. Metapopulation dynamics, abundance, and distribution in a microecosystem. *Science* 281:2045–2047.

Griffith, B., J. M. Scott, J. W. Carpenter, and C. Reed. 1989. Translocation as a species conservation tool: Status and strategy. *Science* 245:477–480.

Haddad, N. M, D. K. Rosenberg, and B. R. Noon. 2000. On experimentation and the study of corridors. *Conservation Biology* 14:1543–1545.

Hale, M. L., P. W. W. Lurz, M. D. F. Shirley, S. Rushton, R. M. Fuller, and K. Wolff. 2001. Impact of landscape management on the genetic structure of red squirrel populations. *Science* 293:2246–2248.

Hanks, J. 2003. Transfrontier Conservation Areas (TFCAs) in Southern Africa: Their role in conserving biodiversity, socioeconomic development, and promoting a culture of peace. *Journal of Sustainable Forestry* 17:121–142.

Hanski, I. 1989. Metapopulation dynamics: Does it help to have more of the same? *Trends in Ecology and Evolution* 4:113–114.

——. 1999. *Metapopulation Ecology*. New York: Oxford University Press.

Hanski, I., and M. Gilpin. 1991. Metapopulation dynamics: Brief history and conceptual domain. *Biological Journal of the Linnaean Society* 42:3–16.

——, eds. 1997. *Metapopulation Biology: Ecology, Genetics, and Evolution.* San Diego: Academic Press.

Harris, L. D., and P. B. Gallagher. 1989. New initiative for wildlife conservation: The need for movement corridors. In Mackintosh 1989, 11–34.

Harris, L. D., and J. Scheck. 1991. From implications to applications: The dispersal corridor principle applied to the conservation of biodiversity. In Saunders and Hobbs 1991b, 189–220.

Harrison, R. L. 1992. Toward a theory of inter-refuge corridor design. *Conservation Biology* 6:293–295.

Harrison, S., and E. Bruna. 1999. Habitat fragmentation and large-scale conservation: What do we know for sure? *Ecography* 22:225–232.

Harvey, A., ed. 1998. *A Sense of Place.* Canmore, Alberta, Canada: The Yellowstone to Yukon Conservation Initiative.

Havlick, D. 2003. Road kill. *Conservation in Practice* 5:30–34.

Hay, K. G. 1990. Greenways and biodiversity. In Hudson 1990, 162–175.

Heilprin, J. 2001. Norton requests ideas for monument changes. *Ashland Daily Tidings.* March 29.

Hellmund, P. C. 1993. A method for ecological greenway design. In Smith and Hellmund 1993, 123–160.

Herrero, S. 1998. Large carnivore conservation. In Harvey 1998, 65–69.

Herrero, J. 2000. Assessing the design and functionality of wildlife movement corridors in the Southern Canmore region. Jacob Herrero Environmental Consulting. http://www.stratalink.com/corridors/wildlife_corridors_report.htm.

Hess, G. R. 1994. Conservation corridors and contagious diseases: A cautionary note. *Conservation Biology* 8:256–262.

Hilsenbeck, R. A., and W. J. Caster. 1999a. Panther Glades, Hendry County, Florida. Conservation and Recreation Lands 2000 Proposal. Tallahassee: The Nature Conservancy.

——. 1999b. Twelvemile Slough, Hendry County, Florida. Conservation and Recreation Lands 2000 Proposal. Tallahassee: The Nature Conservancy.

——. 2000. Boundary Modification for Pinhook Swamp, Baker County, Florida. Conservation and Recreation Lands project. Tallahassee: The Nature Conservancy.

Hobbs, R. J. 1992. The role of corridors in conservation: Solution or bandwagon? *Trends in Ecology and Evolution* 7:38–391.

Hobbs, R. J., and A. J. M. Hopkins. 1991. The role of conservation corridors in a changing climate. In Saunders and Hobbs 1991b, 281–290.

Hoctor, T. S., M. H. Carr, and P. D. Zwick. 2000. Identifying a linked reserve system using a regional landscape approach: The Florida ecological network. *Conservation Biology* 14:984–1000.

Holdridge, L. R. 1978. *Ecología basada en zonas de vida*. San José, Costa Rica: Instituto Interamericano de Ciencias Agrícolas.

Holroyd, G. 1998. Bird conservation in the Yellowstone to Yukon. In Harvey 1998, 71–75.

Hudson, W. E., ed. 1991. *Landscape Linkages and Biodiversity*. Washington, D.C.: Island Press.

Hughes, T. P., A. H. Baird, D. R. Bellwood, M. Card, S. R. Connolly, C. Folke, R. Grosberg, O. Hoegh-Guldberg, J. B. C. Jackson, J. Kleypas, J. M. Lough, P. Marshall, M. Nyström, S. R. Palumbi, J. M. Pandolfi, B. Rosen, and J. Roughgarden. 2003. Climate change, human impacts, and the resilience of coral reefs. *Science* 301:929–933.

IPCC. 2001. *Climate Change, 2001: Synthesis Report*. Ed. R. T. Watson and the Core Writing Team: A Contribution of Working Groups I, II, and III to the Third Assessment Report of the Intergovernmental Panel on Climate Change. Cambridge: Cambridge University Press

Janzen, D. H. 1986. The future of tropical ecology. *Annual Review of Ecology and Systematics* 17:305–324.

Johnsingh, A. J. T., K. Ramesh, Q. Qureshi, A. David, S. P. Goyal, G. S. Rawat, K. Rajapandian, and S. Prasad. 2004. Conservation status of tiger and associated species in the Terai Arc landscape, India. RR-04/001. Wildlife Institute of India, Dehradun.

Johnson, N., A. White, and D. Perrot-Maître. 2001. Developing markets for water services from forests: Issues and lessons for innovators. Washington, D.C.: Forest Trends.

Johnstone, P. 2001. Muskwa–Kechika management plan process update, June 23, 2001. Ministry of Environment, Lands and Parks, Fort St. John, British Columbia, Canada. http://www.gov.bc.ca.

Joshi, A. July 2001. Terai Arc landscape: A new approach for conservation. Developed for the World Wildlife Fund Nepal Programme. Unpublished document.

Kaiser, J. 2001. Bold corridor project confronts political reality. *Science* 293:2196–2199.

Kareiva, P. 1990. Population dynamics in spatially complex environments: Theory and data. *Philosophical Transactions of the Royal Society of London*, series B 330:175–190.

King, P., and L. Reibstein. 1997. An animal superhighway? *Newsweek* 130.4:58.

Kitzhaber, J. 2001. Letter to Department of the Interior Secretary Gail Norton. April 24. Office of Governor John Kitzhaber, Salem, Oregon.

Lahaye, W. S., R. J. Gutiérrez, and H. R. Akçakaya. 1994. Spotted owl metapopulation dynamics in Southern California. *Journal of Animal Ecology* 63:775–785.

Laurance, W. F., and G. B. Williamson. 2001. Positive feedbacks among forest fragmentation, drought, and climate change in the Amazon. *Conservation Biology* 15:1529–1535.

Lawton, R. O., U. S. Nair, R. A. Pielke Sr., and R. M. Welch. 2001. Climatic impact of tropical lowland deforestation on nearby montane cloud forest. *Science* 294:584–587.

Levins, R. 1969. Some demographic and genetic consequences of environmental heterogeneity for biological control. *Bulletin of the Entomological Society America* 15:237–240.

——. 1970. Extinctions. In *Some Mathematical Questions in Biology*, ed. M. Gerstenhaber, 77–101. Providence: American Mathematical Society.

Lindenmayer, D. B., and H. A. Nix. 1993. Ecological principles for the design of wildlife corridors. *Conservation Biology* 7:627–630.

Little, C. E. 1990. *Greenways for America*. Baltimore: John Hopkins University Press.

Loney, B., and R. J. Hobbs. 1991. Management of vegetation corridors: Maintenance, rehabilitation, and establishment. In Saunders and Hobbs 1991b, 299–311.

Lotz, M. A., E. D. Land, and K. G. Johnson. 1997. Evaluation and use of precast wildlife crossings by Florida wildlife. *Proceedings of the Annual Conference of the Southeast Association of Fish and Wildlife Agencies* 51:311–318.

MacArthur, R. H., and E. O. Wilson. 1967. *The Theory of Island Biogeography*. Princeton: Princeton University Press.

Mackintosh, G., ed. 1989. *Preserving Communities and Corridors*. Washington, D.C.: Defenders of Wildlife.

Maehr, D. S., E. D. Land, D. B. Shindle, O. L. Bass, and T. S. Hoctor. 2002. Florida panther dispersal and conservation. *Biological Conservation* 106:187–197.

Mahr, M., M. Soule, and S. Herrero, eds. 1999. Y2Y Science Advisory Forum summary report. Canmore, Alberta, Canada: The Yellowstone to Yukon Conservation Initiative.

Main, M. B., F. M. Roka, and R. F. Noss. 1999. Evaluating the costs of conservation. *Conservation Biology* 13:1262–1272.

Malavasi, E. O., and J. Kellenberg. 2002. Program of payments for ecological services in Costa Rica. http://epp.gsu.edu/pferraro/special/lr_ortiz_kellenberg_ext.pdf *or* http://www.iucn.org/themes/fcp/publications/files/flr_costarica/flr_ortiz_kellenberg_ext.doc.

Mann, C. C., and M. L. Plummer. 1993. The high cost of biodiversity. *Science* 260:1868–1871.

——. 1995. Are wildlife corridors the right path? *Science* 270:1428–1430.

Mayhood, R., R. Ament, R. Walker, and B. Haskins. 1998. Selected fishes of the Yellowstone to Yukon: Distribution and status. In Harvey 1998, 77–91.

McCullough, D. R. 1996. Introduction. In *Metapopulations and Wildlife Conservation*, ed. D. R. McCullough, 1–10. Washington, D.C.: Island Press.

McNeely, J. A., and K. R. Miller, eds. 1984. *National Parks, Conservation, and Development: The Role of Protected Areas in Sustaining Society*. Washington, D.C.: Smithsonian Institution Press.

Medford Mail Tribune. 2001. Brong calls monument a "fine thing." *Medford Mail Tribune.* May 5, 2A.

Merrill, T., and D. J. Mattson. 1998a. Land cover structure of Yellowstone to Yukon. In Harvey 1998, 27–29.

——. 1998b. Defining grizzly bear habitat in the Yellowstone to Yukon. In Harvey 1998, 103–111.

Miller, K., E. Change, and N. Johnson. 2001. *Defining Common Ground for the Mesoamerican Biological Corridor.* Washington, D.C.: World Resources Institute.

Mitchell, J. G. 2001. The Big Open: Going public with public lands. *National Geographic,* August, 2–29.

Nepstad, D. C., A. Verissimo, A. Alencar, C. Nobre, E. Lima, P. Lefebvre, P. Schlesinger, C. Potter, P. Moutinho, E. Mendoza, M. Cochrane, and V. Brooks. 1999. Large-scale impoverishment of Amazonian forests by logging and fire. *Nature* 398:505–508.

Newmark, W. D. 1993. The role and design of wildlife corridors, with examples from Tanzania. *Ambio* 22:500–504.

Noss, R. F. 1987. Corridors in real landscapes: A reply to Simberloff and Cox. *Conservation Biology* 1:159–164.

——. 1991. Effects of edge and internal patchiness on avian habitat use in an old-growth Florida hammock. *Natural Areas Journal* 11:34–37.

——. 1993. Wildlife corridors. In Smith and Hellmund 1993, 43–68.

——. 1995. Maintaining ecological integrity in representative reserve networks. World Wildlife Fund, Canada and USA, Toronto and Washington, D.C.

Noss, R. F., and L. D. Harris. 1986. Nodes, networks, and MUMS: Preserving biodiversity at all scales. *Environmental Management* 10:299–309.

One Country. 2001. In Nepal, a novel project mixes literacy and microfinance to reach thousands. http://www.onecountry.org.

Palomares, F. 2001. Vegetation structure and prey abundance requirements of the Iberian lynx: Implications for the design of reserves and corridors. *Journal of Applied Ecology* 38:9–18.

Partners for Wetlands. 2001. Kinabatangan, a corridor of life: A vision for the Kinabatangan 2020, Lower Kinabatangan Floodplain, Malaysia. World Wildlife Fund, Gland, Switzerland.

Pengelly, I., and C. White. 1998. Fire in the Yellowstone to Yukon. In Harvey 1998, 97–102.

Perrot-Maître, D. 2000. Market-based instruments for watershed management: Case studies around the world. Workshop on Developing Markets for Environmental Services of Forests, Vancouver, British Columbia. Washington, D.C.: Forest Trends.

Porto, M., et al. 1999. A participatory approach to watershed management: The Brazilian system. *Journal of the American Water Resources Association* 35:675–684.

Pulliam, H. R. 1988. Sources, sinks, and population regulation. *American Naturalist* 132:652–661.

Rasker, R., and B. Alexander. 1998. Economic trends in the Yellowstone to Yukon region: A synopsis. In Harvey 1998, 51–56.

Reeves, B. O. K. 1998. Sacred geography: First Nations of the Yellowstone to Yukon. In Harvey 1998, 31–50.

Register-Guard. 2001. Don't shrink the monument. *Register-Guard*. May 1, editorial.

Ressner, J. 2001. A high noon in the west: Logging free-for-all in a forest. *Time*, July 10, 26–28.

Roberts, C. M., and J. P. Hawkins. 2000. *Fully-Protected Marine Reserves: A Guide*. World Wildlife Fund, Washington, D.C.

Rose, C. 2000. Common property, regulatory property, and environmental protection: Comparing common pool resources to tradable environmental allowances. 8th Biennial Conference of the International Association for the Study of Common Property, Indiana University, Bloomington, Indiana.

Rosenburg, D. K., B. R. Noon, and E. C. Meslow. 1997. Biological corridors: Form, function, and efficacy. *Bioscience* 47:677–687.

Saunders, D. A., and R. J. Hobbs. 1991a. The role of corridors in conservation: What do we know and where do we go? In Saunders and Hobbs 1991b, 404–405.

——, eds. 1991b. *Nature Conservation 2: The Role of Corridors*. Chipping Norton, New South Wales, Australia: Surrey Beaty.

Sawyer, M. 1998. Human threats in the Yellowstone to Yukon. In Harvey 1998, 57–60.

Sawyer, M., and D. Mayhood. 1998. Cumulative effects of human activity in the Yellowstone to Yukon. In Harvey 1998, 61–63.

Simberloff, D., and L. G. Abele. 1982. Refuge design and island biogeographic theory: Effects of fragmentation. *American Naturalist* 120:41–50.

Simberloff, D., and J. Cox. 1987. Consequences and costs of conservation corridors. *Conservation Biology* 1:63–71.

Simberloff, D., J. A. Farr, J. Cox, and D. W. Mehlman. 1992. Movement corridors: Conservation bargains or poor investments? *Conservation Biology* 6:493–504.

Singletary, J. G., R. R. L. Hunter, J. Wurth, J. N. Reid, and G. E. Nelson. 2001. Editorial: Leave it alone. *Medford Mail Tribune*.

Smith, D. S., and P. C. Hellmund, eds. 1993. *Ecology of Greenways: Design and Function of Linear Conservation Areas*. Minneapolis: University of Minnesota Press.

Soulé, M. 1987. *Viable Populations for Conservation*. Cambridge: Cambridge University Press.

Soulé, M. E., and M. E. Gilpin. 1991. The theory of wildlife corridor capability. In Saunders and Hobbs 1991b, 3–8.

Stickel, P. F., S. M. Rowe, D. L. Atkin, P. K. Bhatia, K. Denny, and R. J. Caldwell. 2001. Editorial: Let this monument stand. *Oregonian*. May 17, B10.

Strittholt, J. R., R. F. Noss, P. A. Frost, K. Vance-Borland, C. Carroll, and G. Heilman Jr. 1999. A science-based conservation assessment for the Klamath–Siskiyou ecoregion. Earth Design Consultants, Inc., and the Conservation Biology Institute, Corvallis, Oregon.

Thomas, J. W., E. D. Forsman, J. B. Lint, E. C. Meslow, B. R. Noon, and J. Verner. 1990. A conservation strategy for the northern spotted owl: A report to the Interagency Scientific Committee to address the conservation of the northern spotted owl. U.S. Forest Service, U.S. Fish and Wildlife Service, and National Park Service, Washington, D.C.

Thorne, J. F. 1993. Landscape ecology: A foundation for greenway design. In Smith and Hellmund 1993, 23–42.

Turner, M., R. H. Gardner, and R. V. O'Neill. 2001. *Landscape Ecology in Theory and Practice: Pattern and Process.* New York: Springer.

UNDP/UNEP/World Bank/WRI 2000. *World Resources, 2000–2001.*Washington, D.C.: World Resources Institute.

van der Linde, H., J. Oglethorpe, T. Sandwith, D. Snelson, and Y. Tessema. 2001. Beyond boundaries: Transboundary natural resource management in sub-Saharan Africa. Biodiversity Support Program, Washington, D.C.

Vitousek, P. M., H. A. Mooney, J. Lubchenco, and J. M. Mellilo. 1997. Human domination of Earth's ecosystems. *Science* 277:494–499.

Vreugdenhil, A. 2000. *Veluwe2010, een kwaliteitsimpuls!* Provincie Gelderland, Arnhem.

Walker, R., and L. Craighead. 1998. Corridors: Key to wildlife from Yellowstone to Yukon. In Harvey 1998, 113–121.

Weaver, L. C. 1997. Background and potential impacts resulting from construction of a game and livestock proof fence by the Government of Botswana south of the West Caprivi Game Reserve. WWF/LIFE Programme, Namibia. Unpublished report.

Webster, D. 1999. Walking a wildlife highway from Yellowstone to the Yukon. *Smithsonian*, November.

Wells, M., S. Guggenheim, A. Khan, W. Wardojo, and P. Jepson. 1999. Investing in biodiversity: A review of Indonesia's Integrated Conservation and Development Projects. World Bank, East Asia Region, Washington, D.C.

Willcox, L. 1998. The wild heart of North America: A new perspective. In Harvey 1998, 1–3.

Wilson, E. O., and E. O. Willis. 1975. Applied biogeography. In *Ecology and Evolution of Communities*, ed. M. L. Cody and J. M. Diamond, 522–534. Cambridge, Mass.: Belknap Press.

Wood, A., P. Stedman-Edwards, and J. Mang, eds. 2000. *The Root Causes of Biodiversity Loss.* London: World Wildlife Fund and Earthscan Publications Ltd.

WRI. 1992. *World Resources, 1992–1993.* Washington, D.C.: World Resources Institute.

———. 1995. Africa data sampler: A geo-referenced database for all African countries. Washington, D.C.: World Resources Institute.

Wuethrich, B. 2000. When protecting one species hurts another. *Science* 289:383–385.

WWF. 2000. *Stakeholder Collaboration: Building Bridges for Conservation.* Washington, D.C.: World Wildlife Fund.

———. 2001a. Klamath–Siskiyou biodiversity vision and conservation plan. World Wildlife Fund, Ashland, Oregon.

———. 2001b. An ecoregionally-based approach to private lands incentives in the Klamath–Siskiyou. World Wildlife Fund, Ashland, Oregon.

———. 2001c. *Terai Arc: In the Shadow of the Himalayas. A New Paradigm for Wildlife Conservation.* Washington, D.C.: World Wildlife Fund.

Wyden, R. 2001. Letter to Department of the Interior Secretary Gail Norton. May 3. Office of Senator Ron Wyden, Washington, D.C.

Zimmerman, B. L., and R. O. Bierregaard. 1986. Relevance of the equilibrium theory of island biogeography and species-area relations to conservation, with a case from Amazonia. *Journal of Biogeography* 13:133–143.

Index

Abadi Meway, 167–68

Adopt an Acre program, 199n7

Africa: costs of not establishing corridors, 59–60; migration patterns and corridor design, 31; transfrontier conservation areas (TFCAs), 76–77, 78(fig.). *See also specific countries*

agriculture: agroforestry in the Atlantic Forest, 72–73, 91; agroforestry along the Kinabatangan River, 54; agroforestry in the Talamanca–Caribbean region, 104; banana plantations, 98, 102; cocoa plantations, 104; crops destroyed by elephants, 38, 158, 161–62, 173; and landscape connectivity, 22, 104; minimizing impacts in Talamanca–Caribbean region, 101–2. *See also* oil palm plantations

alligator, American (*Alligator mississippiensis*), 110

American dipper (*Cirnclus mexicanus*), 127

AMLD. *See* Golden Lion Tamarin Association

animal–human conflicts: elephant conflicts, 38, 158, 161–62, 173; tiger attacks, 38, 173, 177

animal populations: alternate strategies for species conservation, 24; colonization/immigration, 12–14; and equilibrium theory of island biogeography, 11–13; extinction, 12–14; feeding ecology, 32; inbreeding depression, 24, 106, 197–1; and metapopulation theory, 13–14; protection of keystone, flagship, or umbrella species as corridor objective, 30–33; social organization, 32; specialist vs. generalist species, 52; species-specific responses to corridors, 23, 31. *See also* dispersal ecology; mobility of species

Aracruz Florestal, 88

Atlantic Forest Corridor (Brazil), 72–73, 83–95; background, 84–85; and corridor context, 28; design and implementation, 83, 85–89; environmental services provided by, 63–64; and fire danger, 83, 86, 90–91; governance, 93–94; habitat composition, 84; habitat fragmenta-